T0269522

Feeding Everyone No Matter What
Managing Food Security After Global Catastrophe

Feeding Everyone No Matter What
Managing Food Security After Global Catastrophe

David Denkenberger

Joshua M. Pearce

AMSTERDAM • BOSTON • HEIDELBERG • LONDON
NEW YORK • OXFORD • PARIS • SAN DIEGO
SAN FRANCISCO • SINGAPORE • SYDNEY • TOKYO

Academic Press is an imprint of Elsevier

Academic Press is an imprint of Elsevier
32 Jamestown Road, London NW1 7BY, UK
525 B Street, Suite 1800, San Diego, CA 92101-4495, USA
225 Wyman Street, Waltham, MA 02451, USA
The Boulevard, Langford Lane, Kidlington, Oxford OX5 1GB, UK

Library of Congress Cataloging-in-Publication Data
A catalog record for this book is available from the Library of Congress

British Library Cataloguing in Publication Data
A catalogue record for this book is available from the British Library

ISBN: 978-0-12-804447-6

For information on all Academic Press publications
visit our website at http://store.elsevier.com/

This book has been manufactured using Print On Demand technology. Each copy is produced to order and is limited to black ink. The online version of this book will show color figures where appropriate.

Working together
to grow libraries in
developing countries

www.elsevier.com • www.bookaid.org

DEDICATION

This book is dedicated to our children: Emily, Jerome, Vincent, Audrianna, and Julia – may they never need to use it.

CONTENTS

ACKNOWLEDGMENTS

This book was made possible by the help of a number of people including our friends and family, but also a long list of subject matter experts we consulted to ensure that the solutions proposed are viable given the state of the art in science. These people were all a great help to us and for the creation of the book. Any errors and omissions, however, are ours alone.

We would like to thank our editors Nancy Maragioglio and Carrie Bolger, without their kind help this book would never have existed. We would also like to acknowledge helpful discussions with Joseph Geddes, Carl Shulman, Anders Sandberg, David Fox, Glenn Prestwich, E. Wayne Askew, Robert Andrews, Jennifer Bow, Stuart Armstrong, Kerri Pratt, Sean Brandt, Aaron Socha, Seth Baum, Geoffrey Livesey, Anton Sonnenberg, Owen Cotton-Barratt, Barbara Bentley, Ronald Benner, David Bignell, Aleksandra Walczyńska, Jason English, Peter Paalvast, Jill Stone, Marcel Dicke, Robert Elkin, James Dunn, John Comerford, Tablado Zulima, Jay Stauffer, Jian Shi, Geoffrey Hoy, Karri Bertram, Saravanamuth Vigneswaran, and Ian Rowland.

Dr. David Denkenberger received his bachelor's from Penn State in Engineering Science, his master's from Princeton in Mechanical and Aerospace Engineering, and his doctorate from the University of Colorado at Boulder in the Building Systems Program. His dissertation was on his patent-pending expanded microchannel heat exchanger. He is a research associate at the Global Catastrophic Risk Institute. He received the National Merit Scholarship, the Barry Goldwater Scholarship, the National Science Foundation Graduate Research Fellowship, and is a Penn State distinguished alumnus. He has authored or co-authored over 30 publications and has given over 60 technical presentations.

Dr. Joshua M. Pearce received his Ph.D. in Materials Engineering from the Pennsylvania State University. He then developed the first sustainability program in the Pennsylvania State System of Higher Education as an assistant professor of Physics at Clarion University of Pennsylvania and helped develop the Applied Sustainability graduate engineering program while at Queen's University, Canada. He currently is an associate professor cross-appointed in the Department of Materials Science & Engineering and in the Department of Electrical & Computer Engineering at the Michigan Technological University where he runs the Open Sustainability Technology Research Group. His research concentrates on the use of open source appropriate technology to find collaborative solutions to problems in sustainability and poverty reduction. His research spans areas of

electronic device physics and materials engineering of solar photovoltaic cells, and RepRap 3-D printing, and also includes applied sustainability and energy policy. He has published more than 100 peer-reviewed articles and is the author of the *Open-Source Lab: How to Build Your Own Hardware and Reduce Research Costs.*

CHAPTER *1*

Introduction

1.1 INTRODUCTION TO THE CHALLENGE

The purpose of this book is to answer a simple question: **How do we feed everyone on the planet if we lose mass-scale agriculture due to one or several problems from a long list of global catastrophes?** We think it is important to answer this question because if global agricultural production is dramatically reduced for several years, then mass human starvation is currently likely. That means you, your friends, and your family would all die – probably in a horrible way[1] – and it is all preventable. Such a situation would follow a global crisis such as: super volcano, asteroid or comet impact, nuclear winter (full-scale nuclear war, like between the United States and Russia, accompanied by burning of cities with the smoke going into the upper atmosphere), abrupt climate change (happening in about a decade), super weed (a weed that out-competes crops), super crop pathogen (disease), super bacterium (disrupts beneficial bacteria), or super crop pest (insects, birds, etc.). These risks are generally known as global catastrophic risks, which could destroy or significantly harm civilizations (e.g., Hanson, 2008; Tonn and MacGregor, 2009; Maher and Baum, 2013). Even more serious are the risks that could cause human extinction or permanent loss of civilization, often called existential risks. It is possible that the agricultural risks above could spiral out of control and end up being an existential risk. This would not just affect the current generation, but all future generations. We believe that the food solutions in this book significantly reduce the risk that this will happen. First, we will summarize the severity and probabilities of such scenarios in Chapter 2 covering worldwide crop death. Then, we will explore the more challenging worldwide crop destruction from the temporary yet extreme multiple-year loss of sunlight in Chapter 3.

---

[1] Hollywood has looked into the potential apocalyptic scenarios quite exhaustively – the results are not overly cheery. See, for example, *The Road, The Day, The Book of Eli, Snowpiercer*, or any of the *Mad Max* films.

Feeding Everyone No Matter What. http://dx.doi.org/10.1016/B978-0-12-804447-6.00001-2

The primary historic solution developed over the last several decades for all of these massive problems is simply increased food storage. Modern day survivalists (and their slightly more moderate cousins, the preppers) have embraced this solution.[2] This technique is useful for lesser catastrophes such as personal financial trouble or even widespread economic system collapse, but stocking up the pantry would be very expensive for the severe catastrophes presented in this book, as we show in Chapter 4. In that Chapter, we cover present food supplies and capacity of storing more on a wider scale. Next, we perform a technical evaluation of lone survivalism from a food perspective and for completeness conclude with an evaluation of basic cannibalism mathematics. These "survivalist solutions" are practical only for the most wealthy and thus are not viable for the vast majority of the world's population including the majority of Americans. Using storage on a large scale can actually make the problem worse, as storing up enough food to feed a significant fraction of the population would take a significant amount of time and would increase the price of food, killing additional people due to inadequate global access to affordable food now. Today, thousands starve because they cannot afford access to food, despite the fact that there is more than enough to feed everyone. Any increase in food prices increases this death rate because of the number of food-insecure people living primarily in the developing world.

Humanity is far from doomed, however, even in the worst of these situations – there are solutions and we provide a wide menu to select from here. The solutions that are viable can take a considerable amount of time to ramp up until enough calories are available to feed a significant portion (all) of the population. We call this in-between time stopgap food production "fast food" and cover numerous short-term bridges between existing stored food and technically viable 5-year food solutions in Chapter 5.

In this book, we provide a technical analysis, accurate to within a factor of 10 for all situations (the necessity of broadening the accuracy of the analysis will become clear as we walk through the state-of-knowledge

---

[2] See, for example, any of the long line of books with titles like *"Emergency Food Storage & Survival Handbook: Everything You Need to Know to Keep Your Family Safe in a Crisis," "Just Add Water: Recipes Suitable for Long-Term Food Storage, Natural Disasters and Prepping,"* and *"Prepper's Food Storage: 101 Easy Steps to Affordably Stock a Life-Saving Supply of Food."*

for the necessary long-term future projections). We compare food requirements of all humans for five years with conversion of existing plant matter and fossil fuels to edible food. We quantify the existing supplies of fiber for conversion to food in Chapter 6. Then, we present mechanisms for global-scale food production from the available resources even in the absolute worst-case global catastrophes in Chapter 7. The options include: natural gas-digesting bacteria, extracting food from leaves, and conversion of fiber by enzymes, mushroom, or bacteria growth. Further options involve a two-step process involving partial rotting of fiber by fungi and/or bacteria and feeding them to animals such as beetles, ruminants (cows, deer, etc.), rats, and chickens. We perform an analysis to determine the ramp rates for each option. The good news is the results show that careful planning and global cooperation could ensure that humanity and the bulk of biodiversity could be maintained even in the most extreme circumstances. In fact, there will be even a little dietary variety, as we are sure that some readers who would do it to survive may not be overly excited about eating bacteria, beetles, or rats – every day for five years straight.

Having the knowledge that we could provide for humanity's survival in the worst of catastrophes and then actually doing it are entirely different. Chapter 8 covers the practical matters for ensuring that the solutions provided in Chapter 7 can be acted on including evaluations of energy and water needs, nutrition, taste, biodiversity, and the requirements for adequate cooperation. However, the existence of this book creates a moral hazard. That is, if shortsighted leaders have knowledge of a solution (or in this case many solutions), they may be less motivated to reduce risks we have some control over (and as we see in Chapters 2 and 3, these risks are the most probable). We have convinced ourselves that we have overcome this moral hazard, which is why you are reading this book. Chapter 9 discusses this problem and defends our writing this book. In doing so, we lay out some common sense, rational, and remarkably easy policies that would be in the best interest of everyone to implement immediately to greatly reduce the probability of such global agricultural disasters. All of this said, there are still a lot of unknown primarily technical questions, which must be investigated to convert these solution ideas into full-fledged, hammered-out solutions. Thus, finally, in Chapter 10, we outline the research that is necessary to actualize these solutions and plan for humanity's survival ensuring that we feed everyone no matter what.

REFERENCES

Hanson, R., 2008. Catastrophe, social collapse, and human extinction. In: Bostrom, N., Cirkovic, M.M. (Eds.), Global Catastrophic Risks. Oxford University Press, p. 554.

Tonn, B., MacGregor, D., 2009. A singular chain of events. Futures 41 (10), 706–714.

Maher, Jr., T.M., Baum, S.D., 2013. Adaptation to and recovery from global catastrophe. Sustainability 5 (4), 1461–1479.

CHAPTER 2

Worldwide Crop Death: The Five Crop-Killing Scenarios

2.1 THE FIVE CROP-KILLING SCENARIOS

It is widely assumed that if agricultural production is dramatically reduced over a period of years, this will cause mass human starvation, as seen in Figure 2.1, or even extinction. This could be affected by any of five crop-killing scenarios: (1) abrupt climate change (Valdes, 2011), (2) super weed (Mann, 1999), (3) extirpating (completely destructive) crop pathogen (Dudley and Woodford, 2002), (4) super crop pest (Saigo, 2000), or (5) super bacterium (Church, 2009).

2.2 ABRUPT CLIMATE CHANGE

Of the five crop-killing scenarios – abrupt climate change, or a rapid shift in regional temperatures, is potentially the most serious. Earth has undergone dramatic temperature fluctuations in the past, with 10°C regional changes in one decade (Valdes, 2011). Unfortunately, climate change or global warming has become something of a trigger in the U.S. culture wars. Partially to avoid the quagmire and partially for technical reasons, slow climate change, which occurs over a century or more, will not be emphasized here. Slow climate change allows for adaptations to prevent mass starvation such as relocating crops to more appropriate regions and simply moving people out of regions that cannot provide adequate food. That said, there is overwhelming evidence that greenhouse gas emissions are causing global temperatures to rise. As these emissions continue to destabilize the global climate (Solomon et al., 2007), abrupt climate change could become more intense (Alley et al., 2003).

Abrupt climate change on the decade time scale would have a considerable negative impact on agriculture, because it would be difficult to adapt the global food system this quickly. Today, crops are already having difficulty in some regions, as shown in Figure 2.2, where maize

Feeding Everyone No Matter What. http://dx.doi.org/10.1016/B978-0-12-804447-6.00002-4

Fig. 2.1. A woman, a man, and a child dead from starvation in Russia in 1921. This work is in the public domain in Russia according to Article 6 of Law No. 231-FZ of the Russian Federation of December 18, 2006.

Fig. 2.2. Failed maize crops in Ghana's Upper West Region, which has suffered failed rains and rising temperatures due to climate change. Photograph by Neil Palmer (CIAT) commissioned as part of the Two Degrees Up series of photo stories (CC).

crops failed in Ghana's Upper West Region because of lack of rain and rising temperatures aggravated by climate change. The most serious scenario is considerable global climate change on this time scale. Optimistically, there is no evidence that this has happened (Matthews and Caldeira, 2007). However, there have been several instances of abrupt regional climate change, one approximately 8°C drop in about two decades, and then, several decades later an 8°C rise (GRIP, 1993). Also, there have been six episodes of cooling up to 10°C over several decades in the last 120,000 years (Valdes, 2011). Furthermore, there was a rapid warming episode of 8–16°C over several decades (Matthews and Caldeira, 2007). Therefore, we estimate a probability of 1 in 10,000 per year of natural abrupt climate change. This would not be normally considered a major risk, except for the fact that the climate system is no longer operating under natural conditions. Global carbon dioxide emissions from burning fossil fuels (such as coal, natural gas, and petroleum) have risen to over 36 billion metric tons per year to push the global carbon dioxide (CO_2) concentration above 400 parts per million (ppm) (Le Quéré et al., 2013). We assume that the current anthropogenic (human caused) warming increases the probability of abrupt climate change by an order of magnitude resulting in a probability of 1 in 1000 per year.

2.3 LESSER EVILS – GLOBAL CROP IRRITATING SCENARIOS

Fortunately, the nonclimate altering crises, including super weeds, super crop pathogens, super crop pests, and super bacteria, are generally less serious, and the literature contains fewer quantitative estimates of their impact and probabilities, as they are primarily regional concerns. The current global crop losses from weeds, pathogens and animal pests are 9, 15, and another 11%, respectively (Oerke, 2006).

2.3.1 Super Weeds and Pathogens

Although some so-called "super weeds" have acquired immunity to a herbicide (Kilman, 2010), they are not indestructible invaders. We have numerous examples of weed problems that could become the super weeds of the future. For example, in Figure 2.3, kudzu, a vine from Japan that has become an invasive species in the United States, is seen taking over Atlanta, Georgia. Kudzu is considered a noxious weed (one that is injurious to agricultural and horticultural crops, natural habitats,

Fig. 2.3. Kudzu, a Japanese vine species invasive in the southeast United States, growing in Atlanta, Georgia
(Public Domain).

ecosystems, humans, and/or livestock). Kudzu's environmental and ecological damage simply results from blocking the sunlight to other plants. Biologists call this as "interference competition," meaning kudzu out-competes other species for a resource – in this case, access to solar energy for photosynthesis by growing over them and shading them with its big leaves. The weaker native plants may then die as a result of being buried by a slow motion tidal wave of kudzu.

Perhaps more disturbing, these weeds can collaborate to destroy agriculture. For example, a study published in 2014 found that changes in leaf litter associated with kudzu infestation resulted in changes to decomposition processes and a 28% reduction in stocks of soil carbon (Tamura and Tharayil, 2014). What this means is that kudzu acceleration would be highly likely to enhance the greenhouse effect and accelerate climate change. Kudzu, in particular, at least should not be feared. Although, it is considered a noxious weed by authorities, it is actually food. The leaves, vine tips, flowers, and roots are all edible, although the vines themselves are not. The leaves can be used the same way you would use spinach and eaten raw, chopped up and baked in quiches, cooked like collard greens, or if you need the extra calories: deep-fried.

Kudzu is a surprisingly versatile food source: (1) the young kudzu shoots are tender and taste similar to snow peas; (2) its purple-colored, grape-smelling blossoms can be made into jelly, candy, and syrup; and (3) its large potato-like roots are full of protein, iron, fiber, and useful micro-nutrients, which can be dried and ground into a powder to coat foods for frying or to thicken sauces. Only in a land of ample food would kudzu be considered a pest. (It is interesting to note that the humble dandelion is also considered a weed and an invasive species in the United States; it was introduced to American gardens as a food crop from France as both its roots and leaves are edible.)

However, it is entirely possible that some super weed could develop either naturally or artificially, which could similarly out-compete existing plants including the kudzu and be inedible or even toxic to humans and animals. This may sound scary – but consider that no matter what, they are just weeds and do not represent extinction risk to humanity. They can be controlled by a long list of other herbicides if they manage to overcome one of them. In the absolute worst-case scenario they can be vanquished by even low-tech pulling or tilling the weeds underground. However, there are greater threats of the super weeds releasing toxins or having consider-ably higher photosynthetic efficiency (Mann, 1999), and therefore, over-whelming conventional control. Also, a coordinated terrorist attack using superweeds to destroy crops in large regions could be a global threat.

There is also a rising concern about the susceptibility to pests (next section) and pathogens of crops with limited genetic variability (Saigo, 2000) and the potential for engineered pathogens that target mul-tiple species (Madden and Wheelis, 2003). These threats can be largely mitigated by simply growing a greater variety of crops and humanity as a whole has access to a large selection and variety in all the crop classes. There is already a wide market for so-called heirloom seeds for a wide variety of species and crop scientists are working hard making even more varieties.

2.3.2 Super Pests

The literature defines a "super pest" as an animal (normally in this context the animals discussed are insects) that has acquired pesticide resistance (Saigo, 2000). The adequately available solutions are simply to use other pesticides or natural methods of control to mitigate this risk.

Fig. 2.4. Colorado potato beetle destroying a crop. Photograph released under GNUFDL by Pilise Gábor in 2008.

This is an arms race. For example, the Colorado potato beetle, shown in Figure 2.4, has evolved resistance to over 50 different compounds already belonging to all major insecticide classes. An animal could also become so much more powerful that it overwhelms conventional methods of control. This has happened historically (even Biblically), but only over isolated regions and is highly unlikely to become a global threat because of the great diversity of global ecosystems. For example, although the Colorado beetles are a very serious pest of potatoes and also cause significant damage to tomatoes and eggplants, they do not attack all food crops and are not able to live in all environments. However, again a coordinated terrorist attack using superpests could be a global threat.

2.3.3 Super Bacterium

Perhaps more threatening than the super pests, although much smaller, is the concept of a super bacterium. Super bacteria can be very scary. Consider the humble *Klebsiella pneumoniae* bacteria in Figure 2.5. *K. pneumoniae* is the bacterium first identified with New Delhi Metallo-beta-lactamase-1 (or NDM-1), an enzyme that makes bacteria resistant to a broad range of beta-lactam antibiotics. Although not (yet) a major

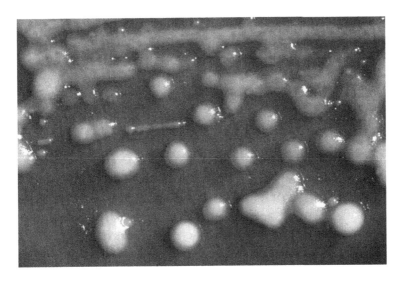

Fig. 2.5. This culture plate cultivated growth of small rod-shaped and anaerobic (without oxygen) Klebsiella pneumoniae *bacteria.* K. pneumoniae *bacteria are commonly found in the human gut, and are often the cause of hospital-acquired infections involving the urinary and pulmonary (lung) systems.* This image is compliments of the U.S. CDC.

problem itself, antibiotic-resistant bacteria could be the next pandemic according to the Centers for Disease Control. Tom Frieden, the CDC's Prevention Director said "Anti-microbial resistance has the potential to harm or kill anyone in the country, undermine modern medicine, to devastate our economy [it already costs us over $20 billion in health care spending a year] and to make our health care system less stable." (Bui, 2014). This scares a lot of people and yet the super bacteria we are concerned with, fortunately, do not yet exist. We consider the super bacterium concept, which refers to a bacterium that disrupts growing conditions for plants by altering soil chemistry or disrupting beneficial bacteria (Church, 2009). There are no credible examples of bacteria that are not pathogens that disrupt crops now, but they could be engineered.

No matter how grave the potential harm for a given region, again, it is unlikely that a single organism represents a global threat, as each would probably not be able to compete in all of the agricultural climates. However, a coordinated terrorist attack of multiple organisms could represent a global threat and should not be completely dismissed.

Barring deliberate global spread bioterrorism with the purpose of eliminating the human species, the spread of any of these super organisms would likely be similar to the spread of invasive species. Thus, there would be enough warning to solve these problems regionally before the global food supply was seriously affected.

2.4 SERIOUS PROBLEMS THAT DO NOT THREATEN GLOBAL FOOD SUPPLY

We have a lot of legitimate regional problems that can threaten the regional food supply. Unfortunately, these regional problems are often blown out of proportion globally. These regional threats include problems such as loss of fisheries, species extinction, loss of bees, loss of unsustainable irrigation, loss of artificial pesticides and fertilizers, loss of topsoil, salinization (turning salty) of soil, desertification, loss of stratospheric ozone, water pollution, and other resource exhaustion issues. These are less serious because of a smaller and/or slower impact on global food production.

For example, losing all the fish in the ocean would be a tragedy and sounds really scary. Yet from a human food perspective, it would not be that big a problem. Global fisheries make up less than 3% of human calories; therefore, even if they all collapsed it would not be a global food catastrophe.

Biodiversity supplies ecosystem services to agriculture such as pollination, pest control, soil fertility, and climate stability (Ehrlich and Ehrlich, 2013). This is important and valuable for humanity, but there are also well-known alternatives and interventions available to us, and therefore, there does not appear to be a credible route to agricultural supply collapse. That said, loss of wild species is important for the species' intrinsic value and we do not mean to diminish that value by ignoring it. We remind the reader that here we are only concerned with preventing global food catastrophes.

If you ingest a lot of mainstream media, you are forgiven for wondering at this point "What about bees?" For example *Wired* ran an article discussing the plight of America's honeybees: the collapsed colonies and dying hives, threatening pollination services to crops and the future of a much-beloved insect (Keim, 2014). However, *Wired* also warned that many wild pollinators are also threatened including thousands of species including bees and butterflies and moths. An ecologist warns that nearly 90% of the world's flowering species require insects or other animals for pollination (Keim, 2014). After all there is a common quote (Often misattributed to Einstein of all people!) "If the bee disappears from the surface of the earth, man would have no more than four years to live." This greatly overstates the value of pollinators for human food production. The total loss in agricultural production for a loss of all animal pollinators is only 3–8% (Aizen et al., 2009). Though this percentage is growing, not all animal pollinators are bees. Therefore, the complete loss of bees, although extremely unfortunate, would not be an agricultural disaster of the magnitude focused on in this book.

Similar to the sensationalized bee death stories, there is a growing concern that we are running out of water that may lead to "water wars" (Wolf, 1999). This can be a serious regional or local problem – but not globally. Total water consumption for food production including rain-fed crops and grazing is ~14,000 billion tons (Gt) per year (Postel, 1998). Irrigation consumption is 1200 Gt per year, about 1/3 of this being unsustainable (Millennium Ecosystem Assessment, 2005). Therefore, losing all unsustainable irrigation would only be a ~3% loss in food production. Furthermore, it would take decades for the unsustainable irrigation to be exhausted.

Readers of books about "Peak Oil" may be concerned that the continued rise in the cost and decline in the availability of fossil fuels may make pesticide and fertilizer production impossible. This is aggravated by how articles cover the fact that huge inputs of fossil fuels are currently used to provide our food supply (Green, 1978; Coley et al., 1998). The key word in the previous sentence is "currently." There are other ways, and as fossil fuels, in general, are relegated to the dustbin of history by the strong growth of renewable energy technologies, these ways will become more common. First, it should be pointed out that not all pesticides are petroleum-based, and it is possible to synthesize petroleum substitutes from renewable hydrogen (via electrolysis [using electricity to split water into hydrogen and oxygen]) (Olah et al., 2009). The dominant production method for nitrogen fertilizer is combining hydrogen with atmospheric nitrogen (Erisman et al., 2008), and renewable energy can provide both the energy and the hydrogen (Pearce, 2002). There are also a variety of solutions for the supply of phosphorus and other mineral fertilizers. Many solutions are similar to the solutions to a food shortage, such as reducing food waste, food fed to animals, land planted with non-food items like cotton, and land planted with lower caloric efficiency foods like coffee. Other solutions involve recycling, such as from wood-burning ash. Additional mining options include landfills (which could involve extracting the paper, wood, etc., converting the energy, and reclaiming the minerals) and even mining commonly occurring rock (Tietenberg, 2000). There are many other mineral inputs to agriculture, e.g., copper for motors. However, options for dealing with a shortage include recycling, substituting, mining landfills, and mining commonly occurring rock.

Finally, loss of topsoil, salinization of soil, desertification, current loss of stratospheric ozone, and water pollution are all relatively slow processes. Humanity as a whole should be more than nimble enough to stop all of these problems from becoming global catastrophes, although the readers will be forgiven for their skepticism on this assumption given that any of these problems exist at all.

Overall, these problems do deserve study in order to be eliminated, but the solutions are more straightforward than for the crises that are the focus of this book, particularly the three most dangerous scenarios discussed in the next Chapter.

2.5 FOOD SPOILAGE

Super pests or microbes could cause considerably more food spoilage than the current pest/microbe scenario. Though many of the solutions presented in this book could ameliorate an increased food spoilage problem, there are more direct solutions that are outside the focus of this book.

There are already bacteria and fungi that can cause food spoilage in dried food at elevated water activity (equilibrated relative humidity) (75% and 61%, respectively) (Grant, 2004). An engineered organism could conceivably cause food spoilage at even lower water activity. One solution for grain would be drying food to lower water levels than happen spontaneously, and then enclosing the food (instead of using porous, grain-storage structures). These organisms could also cause spoilage in salted and sugared food. Because there is limited solubility of salt and sugar, it would not be easy to increase concentrations in the liquid form, so a resolution would be drying so that there is no free water (the only water remaining is chemically bound to the food). Engineered food-spoiling organisms might also be more tolerant of high temperature, thus foiling conventional canning. A solution to this problem would be using higher canning temperatures, but this would require higher pressures and possibly different materials. Engineered food-spoiling organisms might be able to grow at temperatures in typical freezers. One solution would be using lower-temperature freezers. Fermentation preservation of food produces alcohol or acid that discourages other microbes; a super microbe could overcome these impediments. However, in most cases, additional impediments are present, such as sealing of alcoholic beverages and refrigeration of cheese, so this would not be a major problem. It is also possible that a super microbe would spoil fresh food more quickly (refrigerated or not). However, it is likely that this type of microbe would also overwhelm the plant defenses when the plant is alive, so we would classify this as a pathogen. Other methods of preservation preserve relatively minor amounts of food, so their loss would not constitute a crisis. General solutions include switching to different preservation methods and consuming food sooner after harvest (e.g., by producing food more locally).

REFERENCES

Aizen, M.A., Garibaldi, L.A., Cunningham, S.A., Klein, A.M., 2009. How much does agriculture depend on pollinators? Lessons from long-term trends in crop production. Ann. Bot. 103, 1579–1588.

Alley, R.B., Marotzke, J., Nordhaus, W.D., Overpeck, J.T., Peteet, D.M., Pielke Jr, R.A., Pierrehumbert, R.T., Rhines, P.B., Stocker, T.F., Talley, L.D., Wallace, J.M., 2003. Abrupt climate change. Science 299, 2005–2010.

Bui, Hoai-Tran, USATODAY 5:18 p.m. EDT http://www.usatoday.com/story/news/nation/2014/07/22/antibiotic-resistance-bacteria-drugs-cdc-lab-safety-mers-anthrax/13005415/(accessed 22.07.2014).

Church, G., 2009. Safeguarding biology. Seed 20, 84–86.

Coley, D.A., Goodliffe, E., Macdiarmid, J., 1998. The embodied energy of food: The role of diet. Energy policy 26 (6), 455–460.

Dudley, J.P., Woodford, M.H., 2002. Bioweapons, biodiversity, and ecocide: Potential effects of biological weapons on biological diversity. Bioscience 52, 583 (doi:10.1641/0006-3568 (2002)052[0583:BBAEPE]2.0.CO;2).

Ehrlich, P.R., Ehrlich, A.H., 2013. Can a collapse of global civilization be avoided? Proc. R. Soc. London, Ser. B 280 no. 1754, p. 20122845.

Erisman, J.W., Sutton, M.A., Galloway, J., Klimont, Z., Winiwarter, W., 2008. How a century of ammonia synthesis changed the world. Nat. Geosci. 1, 636–639.

Grant, W.D., 2004. Life at low water activity. Phil. Trans. R. Soc. Lond. B 359, 1249–1267.

Green, M.B., 1978. Eating Oil. Energy Use in Food Production. Westview Press Inc. Boulder, CO.

Greenland Ice Core Project (GRIP) Members, 1993. Climate instability during the last interglacial period recorded in the GRIP ice core. Nature 364, 203–207.

Kilman, S., 2010. Superweed outbreak triggers arms race. Wall Str. J. http://online.wsj.com/articles/SB10001424052748704025304575284390777746822.

Keim, B. 2014. Beyond Honeybees: Now Wild Bees and Butterflies May Be in Trouble. Wired. http://www.wired.com/2014/05/wild-bee-and-butterfly-declines/.

Le Quéré, C., Peters, G.P., Andres, R.J., Andrew, R.M., Boden, T., Ciais, P., Friedlingstein, P., Houghton, R.A., Marland, G., Moriarty, R., Sitch, S., Tans, P., Arneth, A., Arvanitis, A., Bakker, D.C.E., Bopp, L., Canadell, J.G., Chini, L.P., Doney, S.C., Harper, A., Harris, I., House, J.I., Jain, A.K., Jones, S.D., Kato, E., Keeling, R.F., Klein Goldewijk, K., Körtzinger, A., Koven, C., Lefèvre, N., Omar, A., Ono, T., Park, G.-H., Pfeil, B., Poulter, B., Raupach, M.R., Regnier, P., Rödenbeck, C., Saito, S., Schwinger, J., Segschneider, J., Stocker, B.D., Tilbrook, B., van Heuven, S., Viovy, N., Wanninkhof, R., Wiltshire, A., Zaehle, S., 2013. Global Carbon Budget 2013, Earth System Science Data Discussion, 6, 689–760. (doi: 10.5194/essdd-6-689-2013)

Madden, L.V., Wheelis, M., 2003. The threat of plant pathogens as weapons against U.S. crops. Annu. Rev. Phytopathol 41. 155–176. (doi:10.1146/annurev.phyto.41.121902.102839)

Mann, C.C., 1999. Genetic engineers aim to soup up crop photosynthesis. Science 283, 314–316.

Matthews, H.D., Caldeira, K., 2007. Transient climate–carbon simulations of planetary geoengineering. Proc. Natl. Acad. Sci. USA 104, 9949–9954.

Millennium Ecosystem Assessment, 2005. Ecosystems and Human Well-Being: Current State and Trends. United States, Island Press.

Oerke, E.C., 2006. Crop losses to pests. J. Agric. Sci. 144, 31–43.

Olah, G., Goeppert, A., Prakash, G.K.S., 2009. Chemical recycling of carbon dioxide to methanol and dimethyl ether: From greenhouse gas to renewable, environmentally carbon neutral fuels and synthetic hydrocarbons. J. Org. Chem. Persp. 74 (2), 487–498.

Pearce, J.M., 2002. Photovoltaics—a path to sustainable futures. Futures 34 (7), 663–674.

Postel, S.L., 1998. Water for food production: Will there be enough in 2025? Bioscience 48, 629–637.

Saigo, H., 2000. Agricultural biotechnology and the negotiation of the biosafety protocol. Georgetown Int. Environ. Law Rev. 12 (3), 779–816.

Solomon, S., Qin, D., Manning, M., Chen, Z., Marquis, M., Averyt, K.B., Tignor, M., Miller, H.L., 2007. Climate Change 2007: The Physical Science Basis Summary for Policymakers. Cambridge University Press, Cambridge, United Kingdom and New York, NY, USA, 1–18.

Tamura, Mioko, Tharayil, Nishanth, 2014. Plant litter chemistry and microbial priming regulate the accrual, composition and stability of soil carbon in invaded ecosystems. New Phytol. 203 (1), 110–124.

Tietenberg, T.H., 2000. Environmental and Natural Resource Economics, fifth ed. Addison-Wesley, Reading, MA.

Valdes, P., 2011. Built for stability. Nat. Geosci. 4, 414–416. (doi:10.1038/ngeo1200)

Wolf, A.T., 1999. "Water wars" and water reality: conflict and cooperation along international waterways in: Environmental Change, Adaptation, and Security. Springer Netherlands, 251–265.

CHAPTER 3

No Sun: Three Sunlight-Killing Scenarios

3.1 THREE SUNLIGHT-KILLING SCENARIOS

The sun (as shown in Figure 3.1), our sun, is incredibly powerful – and is what makes human life possible on the Earth. The total solar radiation falling on the earth is 1.2×10^{14} kW, which is roughly 10,000 times current world energy consumption (Pearce, 2002). Solar photovoltaic (PV) technology, which converts sunlight directly into electricity, offers the most promising method to maintain our energy-intensive standard of living while enabling a long-term sustainable civilization for all of the world's people (Pearce, 2002). Normally, the argument against using PV is that it is uneconomic. Yet, given the state-of-the-art in PV and favorable financing terms, it is clear that PV has already obtained grid parity[1] in specific locations. As installed costs continue to decline, grid electricity prices continue to escalate, and solar industry experience increases, PV will become an increasingly economically advantageous source of electricity over expanding geographical regions (Branker et al., 2011). Primarily because of government policy support solar energy use has been growing rapidly, but it still makes up only a small sliver of the energy pie. The penetration of solar PV technologies is 0.17% of the electricity supply (IEA, 2012), and electricity is 38% of primary (input) energy (Sims et al., 2003). There are other ways to use the sun for energy such as domestic solar hot water, passive solar building heating, and intentional day lighting (substitution of electric light with natural light). However, these solar technologies also have a small penetration. From an engineering perspective, this is sad because passive solar technologies were well developed even in the times of ancient Greeks. Solar energy provides a small amount of heat and light in buildings that the designers did not specifically intend. Other uses of solar energy, such as concentrating

---

[1] Grid parity occurs when an alternative energy source can generate electricity at a levelized cost of electricity (LCOE) that is less than or equal to the price of conventional grid electricity.

Feeding Everyone No Matter What. http://dx.doi.org/10.1016/B978-0-12-804447-6.00003-6

Fig. 3.1. Giant prominence on the sun. Image provided by the NASA/SDO/AIA/Goddard Space Flight Center (Public Domain).

solar power, are relatively small. Therefore, the total contribution of solar energy currently to primary energy is order of magnitude 1%.

But what would happen if we partially lost access to sunlight? The ramifications would go far beyond hurting the economic viability of PV systems and the other small applications of using solar power for energy. The primary way that humanity takes advantage of solar energy today is through the relatively inefficient process of photosynthesis – we use it to grow crops, trees, etc. The land requirement to grow current food and forest products is approximately 5.7 billion hectares (ha) (White, 2007), so with 200 W/m^2 average solar energy (Pearce, 2008) this is ~11,000 trillion watts (TW). For perspective, note that global conventional primary energy consumption is approximately 17 TW (IEA, 2012). Here, we will look at three potential scenarios that could partially obscure the sun and thus render all conventional agriculture impossible: (1) asteroid or comet impact, (2) a super volcano, or (3) nuclear winter.

An asteroid (as shown in Figure 3.2) or meteor (as shown in Figure 3.3) larger than about 1 km diameter striking the earth has the potential to block the sun globally (Bostrom and Cirkovic, 2008). Particles in the troposphere (the lower layer of the atmosphere below about 10 km [6.2 miles]) would quickly rain out, but particles in the stratosphere (the layer above the troposphere), especially sulfate, could remain there

Fig. 3.2. An artist's depiction of watery asteroid. Image provided by NASA, ESA, M.A. Garlick (space-art. co.uk), University of Warwick, and University of Cambridge (CC).

for years (Bostrom and Cirkovic, 2008). There is a significant mass of historical evidence that major asteroid impacts in the past have occurred on Earth. For example, an enormous 3.26-billion-year-old 37 km wide asteroid boiled the top of Earth's oceans, turned the sky red hot, and generated a half-hour-long earthquake that shook the entire planet (Lowe et al., 2003; Sleep and Lowe, 2014). That asteroid was at least 37 times larger diameter than one that would blot out the sun long enough to possibly destroy civilization. The dinosaur-killing asteroid was about

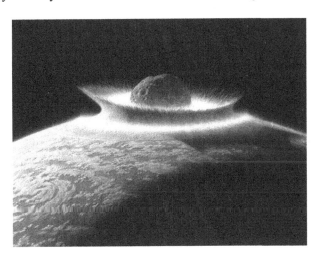

Fig. 3.3. An artist's depiction of a cataclysmic meteor impact compliments of NASA (Public Domain).

Fig. 3.4. An artist's depiction of what the Toba eruption might have looked like from approximately 26 miles above Pulau Simeulue (CC).

a quarter of the diameter at 10 km (or only 10 times larger diameter than what is needed to dim the sun for five years) (Mann, 2014). After five years, the effect would only be ~10% as much as the peak.

Similarly, a supervolcano would reduce the potential of agricultural production from solar photosynthesis. The Toba super volcanic eruption approximately 70,000 years ago as shown in Figure 3.4 may have nearly caused the extinction of humans (Bostrom and Cirkovic, 2008). Again, this is caused primarily by sulfate particles that could block the sun for years.

Solar blocking is not limited to natural causes, as "nuclear winter" is also possible. Nuclear winter, as depicted by an artist in Figure 3.5, refers to the scenario involving widespread nuclear war and burning of cities that release soot (smoke) into the stratosphere (Bostrom and Cirkovic, 2008). The high stratospheric temperatures produced by soot absorbing solar radiation would create near-global ozone hole conditions even for a regional nuclear conflict (Mills et al., 2008). For an all-out nuclear war, the situation would be much worse. The probability of this happening is uncomfortably high. The Bulletin of Atomic Scientists maintains a "Doomsday Clock" shown in Figure 3.6. The Doomsday Clock is an internationally recognized design of the analog clock that conveys how close humanity is to destroying our civilization with dangerous technologies of our own making (BAS, 2014). Originally, it only followed the probability of nuclear war, but has been expanded to

Fig. 3.5. An artist's depiction of nuclear winter. Available for reuse: http://www.1zoom.ru/%D0%98%D0%B3%
D1%80%D1%8B/%D0%BE%D0%B1%D0%BE%D0%B8/350878/z741.8/

*Fig. 3.6. The Bulletin of Atomic Scientist's Doomsday Clock superimposed over a photograph of a 23 kiloton
nuclear test taken by the Federal Government of the United States* (Public Domain).

other technical doomsday scenarios such as climate change. As can be seen in Figure 3.6, there is only five figurative minutes left to midnight (the end of humanity). Five minutes is not a lot of time in the grand experiment of humanity. There have been a staggering 125,000 nuclear warheads built since 1945 (with 97% of American and Russian/USSR manufacture) (Norris and Kristensen, 2010). The nine nations (USA, Russia, UK, France, China, India, Pakistan, Israel, and North Korea) with nuclear weapons now possess more than 10,000 nuclear warheads in their military stockpiles (Norris and Kristensen, 2010). Only a tiny fraction of these (about 50) would be needed to generate significant climate disruption (Robock, 2011).

The duration of the crisis depends on the type of atmospheric particles. For soot from nuclear winter, the temperature reduction decays to 1/e (~40%) of the peak after about a decade because the sun lofts the dark particles (Robock et al., 2007). For the super volcano and extraterrestrial impacts, assuming there are not large amounts of forests burned, the primary particle is sulfate, which decays to 1/e in ~2 years (English et al., 2013).

If we consider the more serious threat from Chapter 2 (abrupt climate change) and the three scenarios here, the probability, severity, and intensity of disasters are summarized in Table 3.1.

There are several interesting conclusions we draw from Table 3.1. First, eliminating sunlight is obviously the best way to get agricultural losses to 100% quickly, abrupt climate change seems slow and far less horrible[2] by comparison. However, recovering from climate change is much more difficult than simply letting the dust fall out of the atmosphere in the three scenarios for this Chapter – and thus takes 10 times longer. Fortunately, the disasters that we have the least control over (comet or asteroid impact and super volcano) also have the lowest probability of actually occurring. The two disasters which we do have the most (all of the) control over, abrupt climate change and nuclear winter, are the most probable. Therefore, it makes the most sense to try to simply solve these problems before they happen and we will discuss this in Chapter 10.

---

[2] It should be noted that losing 10% of global agricultural output may kill millions of people from starvation. However, by comparison to the other scenarios, it is much better than billions.

Table 3.1 Probability, Severity, and Intensity of Disasters					
Disaster	Probability (per year)	Maximum Food Production Loss (%)	Agricultural Loss Velocity (%/yr)	Time to Recovery (yr)	Qualitative Intensity
Abrupt climate change	1 in 1,000[a]	10[e]	1[i]	100[m]	Medium
>1 km asteroid or comet	1 in 1,000,000[b]	100[f]	100[j]	3-10[n]	High
Super volcano	1 in 100,000[c]	100[g]	100[k]	3[o]	High
Nuclear winter	1 in 1,000[d]	100[h]	100[l]	10[p]	Very High

[a] Abrupt climate change on the decadal scale would have a considerable negative impact on agriculture because it would be difficult to adapt this quickly. The most serious scenario is considerable global climate change on this timescale, but there is no evidence that this has happened (Matthews & Caldeira, 2007). However, there have been several instances of abrupt regional climate change, one approximately 8°C drop in about two decades, and then several decades later a 8°C rise (GRIP, 1993). Also, there have been six episodes of cooling up to 10°C over several decades in the last 120,000 years (Valdes, 2011). Furthermore, there was a rapid warming episode of 8 to 16°C over several decades (Matthews & Caldeira, 2007). Therefore, we estimate a probability of 1 in 10,000/yr of natural abrupt climate change. We assume that current anthropogenic warming increases the probability of abrupt climate change by an order of magnitude.

[b] (Bostrom & Cirkovic, 2008).

[c] There is uncertainty, but this is the right order of magnitude (Mason, Pyle, & Oppenheimer, 2004).

[d] One estimate of the chance of full-scale nuclear war is ~1%/yr and it uses conditional probabilities (Hellman, 2008). Another estimate is ~1%/yr, based on analysis of the possibility of U.S. or Russian launches in response to false alarms in their early warning systems (Barrett, Baum, & Hostetler, 2013). Then we conservatively estimate that there is only a 1 in 10 chance of full-scale nuclear war causing nuclear winter.

[e] The regional abrupt climate change could wipe out agriculture in one region, which would have order of magnitude 10% agricultural supply reduction.

[f] (Bostrom and Cirkovic, 2008).

[g] There was a ~50% reduction in primary productivity regionally for the Toba supervolcano (Timmreck et al., 2012), so a larger supervolcano could be order 100% crop reductions globally.

[h] There would be much shorter growing seasons and ozone loss (Robock et al., 2007), so order 100%.

[i] The shortest time was ~1 decade (GRIP, 1993).

[j] ~1 year (similar mechanism to supervolcano and nuclear winter).

[k] ~1 year (English et al., 2013).

[l] ~1 year (Robock et al., 2007).

[m] Some changes had recovery time is less than one century (GRIP, 1993), and some required longer (Valdes, 2011), so we assume one century.

[n] This depends on whether the aerosols are sulfate or black-carbon dominated.

[o] Sulfate aerosol (English et al., 2013).

[p] Black carbon (soot) aerosol (Robock et al., 2007).

REFERENCES

Barrett, A.M., Baum, S.D., Hostetler, K.R., 2013. Analyzing and reducing the risks of inadvertent nuclear war between the United States and Russia. Sci. Global Secur. 21, 106–133.

Bostrom, N., Cirkovic, M.M. (Eds.), 2008. Global Catastrophic Risks. Oxford University Press, New York.

Branker, K., Pathak, M.J.M., Pearce, J.M., 2011. A review of solar photovoltaic levelized cost of electricity. Renew. Sustain. Energy Rev. 15, 4470–4482.

English, J.M., Toon, O.B., Mills, M.J., 2013. Microphysical simulations of large volcanic eruptions: Pinatubo and Toba. J. Geophys. Res. Atmos. 118, 1880–1895.

GRIP, Greenland Ice Core Project Members, 1993. Climate instability during the last interglacial period recorded in the GRIP ice core. Nature 364, 203–207.

Hellman, M.E., 2008. Risk analysis of nuclear deterrence. The Bent of Tau Beta Pi.

IEA, 2012. World Energy Outlook 2012. International Energy Agency.

Lowe, D.R., Byerly, G.R., Kyte, F.T., Shukolyukov, A., Asaro, F., Krull, A., 2003. Spherule beds 3.47–3.24 billion years old in the Barberton Greenstone Belt. South Africa: a record of large meteorite impacts and their influence on early crustal and biological evolution. Astrobiology 3 (1), 7–48.

Mann, A., 2014. Ancient Asteroid Boiled Oceans, Burned the Sky, and Shook Earth for a Half-Hour. *Wired*. http://www.wired.com/2014/04/giant-asteroid-impact/.

Mason, B., Pyle, D., Oppenheimer, C., 2004. The size and frequency of the largest explosive eruptions on Earth. Bull. Volcanol. 66, 735–748.

Matthews, H.D., Caldeira, K., 2007. Transient climate–carbon simulations of planetary. Geoengineering 104, 9949–9954.

Mills, M.J., Toon, O.B., Turco, R.P., Kinnison, D.E., Garcia, R.R., 2008. Massive global ozone loss predicted following regional nuclear conflict. Proc. Natl. Acad. Sci. USA 105, 5307–5312.

Norris, R.S., Kristensen, H.M., 2010. Global nuclear weapons inventories, 1945–2010. Bull. Atom. Scient. 66 (4), 77–83.

Pearce, J., 2002. Photovoltaics – A path to sustainable futures. Futures 34 (7), 663–674.

Pearce, J.M., 2008. Thermodynamic limitations to nuclear energy deployment as a greenhouse gas mitigation technology. Int. J. Nucl. Governance, Econ. Ecology 2, 113–130.

Robock, A., Oman, L., Stenchikov, G.L., 2007. Nuclear winter revisited with a modern climate model and current nuclear arsenals: Still catastrophic consequences. J. Geophys. Res. Atmos. 112, 1984–2012.

Robock, A., 2011. Climatic Consequences of Nuclear Conflict. In AGU Fall Meeting Abstracts 1, 07.

Sims, R., Rogner, H.-H., Gregory, K., 2003. Carbon emission and mitigation cost comparisons between fossil fuel, nuclear and renewable energy resources for electricity generation. Energ. Pol. 31 (13), 1315–1326.

Sleep, N.H., Lowe, D.R., 2014. Physics of crustal fracturing and chert dike formation triggered by asteroid impact, ~3.26 Ga, Barberton greenstone belt, South Africa. Geochem. Geophys. Geosyst. 15 (4), 1054–1070.

The Bulletin of Atomic Scientist's Doomsday Clock Overview. 2014. Available: http://thebulletin.org/overview.

Timmreck, C., Graf, H.-F., Zanchettin, D., Hagemann, S., Kleinen, T., Krüger, K., 2012. Climate response to the Toba super-eruption: Regional changes. Quat. Int. 258, 30–44.

Valdes, P., 2011. Built for stability. Nat. Geosci. 4, 414–416. (doi:10.1038/ngeo1200)

White, T.J., 2007. Sharing resources: The global distribution of the ecological footprint. Ecol. Econ. 64, 402–410.

CHAPTER 4

Food Storage, Food Conservation, and Cannibalism

4.1 REDUCTION OF PRE-HARVEST LOSSES

Before we get into the details of providing new food for a long-term mega-disaster, let us first consider how far down the difficulty ladder we can push the problem by simply not wasting food. The starting point for this analysis is determining the current food production. Grain production is ~2.3 Gt/yr (USDA, 2013), and grains are ~29% total of fiber and moisture (USDA, 2006; Iowa State University Extension, 2006), so this is ~1.6 Gt/yr dry carbohydrate equivalent. Grains make up about half of the calories produced (Meadows et al., 2004); therefore, the total production is ~3.3 Gt dry/yr. Current pre-harvest losses from pests, weeds, etc., are 35%, and best regional practices can cut these in half over time (Oerke, 2006). The crisis in question would likely provide enough incentive for people to move to the best regional practices quickly. The harvest would increase from 65% of no-loss to 82.5% of no-loss, or a 27% increase. Mechanisms for achieving this quickly include ramping up pesticide production and using more labor-intensive techniques to reduce pesticide application in certain areas while maintaining productivity. Farmers could also employ nonpesticide methods of reducing losses, such as biological control where natural predators of pests and weeds are encouraged.

4.2 INCREASED FOOD SUPPLY FOR MODERATE DISASTERS

For one of the less-critical disasters (like those in Chapter 2), another source of increased food supply to make up for the losses is increasing the growth potential with increased irrigation, fertilizers, and high-yield varieties. It would be difficult to implement large irrigation projects, like large dams, rapidly. However, humanity could use water from irrigation more efficiently in a number of ways, including reducing the losses in storage and conveyance, such as covering aqueducts and fixing

Feeding Everyone No Matter What. http://dx.doi.org/10.1016/B978-0-12-804447-6.00004-8

leaks (Wallace, 2000). Though it would be difficult to build soil quality rapidly, farmers could use surface soil treatment to reduce evaporation from the soil and reduce runoff. Humanity could multiply high-yield varieties quickly. Furthermore, industry could ramp up production of fertilizers somewhat, and farmers could utilize animal wastes that they currently do not use as fertilizer. A yield increase of 58% is most of what is possible with mostly fertilizer and irrigation improvements (Foley et al., 2011). We assume half of the potential increase to be achievable quickly, or a 29% increase.

An additional step is reducing the postharvest losses, which are currently ~35% (Godfray et al., 2010). This includes better harvest, storage, and transportation systems. The harvest losses with mechanization are substantial, but the labor savings offset this economically (Kantor et al., 1997). Therefore, in a crisis, farmers could use more labor-intensive harvesting techniques to reduce losses. Loss reduction also includes reduction of waste in retailing and household use, which people can implement even more rapidly. Therefore, we assume a reduction in these types of wastes of a factor of two. This waste is nontrivial. Consider that in the United States an amazing 40% of food that is grown is uneaten, as shown in Figure 4.1, which has a value of $165 billion a year (Gunders, 2012). As Figure 4.1 shows the bulk of the waste is actually at the consumer level in stage 5 of the supply chain for the food system with a third of the seafood and more than a quarter of the grain and fruits and vegetables discarded. This uneaten food ends up rotting in landfills as the single largest component of U.S. municipal solid waste where it accounts for a large portion of U.S. methane emissions (Gunders, 2012), which in turn aggravates climate change, and thus, increasing the likelihood of one of the disasters we are considering. Reducing these food losses would provide more than enough food for the food insecure in the United States (see Information Box 4.1). On the other hand, it is true that the fact that we waste more food now gives us more leeway should a disaster occur. Further analysis is required to determine the effects of reducing food waste overall.

The final step considered quantitatively is humanity reducing the amount of losses of edible food fed to livestock and pets and used to produce biofuels. From a human food perspective these losses currently are extreme. Much has already been written about the extremely inefficient conversion of plant-based calories and protein to animal

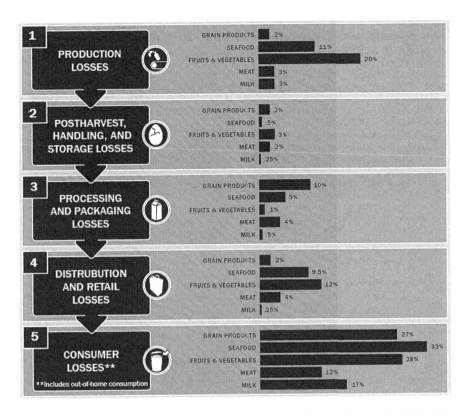

Fig. 4.1. Food losses at each stage of the supply chain for the food system calculated collectively for USA, Canada, Australia, and New Zealand. Adapted from the NRDC and FAO (Gunders, 2012).

protein (Lappé, 2010) and the effects of biofuel subsidies on food prices effectively starving the world's poor (Runge and Senauer, 2007; Brown, 2008). We will not belabor the points, but instead assume the crisis will make these food losses negligible, especially because farmers would still produce some animal products on land that is unfit for human food production directly and on cellulosic food residues. Therefore, starting with 3.3 Gt/yr equivalent of harvest, the crisis would increase harvest 27% due to reduced preharvest losses and another multiplicative 29% due to increased growth potential, yielding ~5.4 Gt/yr. Then, with crisis postharvest losses of 18% and no edible food going to animals or biofuels, this would be 4.4 Gt of human consumption. This is >200% more than the consumption requirement by humans (1.3 Gt dry/yr without losses). It should be noted clearly that reducing food waste for nonsevere crisis can present a solution in

Information Box 4.1

In the report Household Food Security in the United States in 2012, Alisha Coleman-Jensen, Mark Nord, and Anita Singh found that 14.5% of U.S. households (17.6 million households) were designated by the USDA as "food insecure" (2013). This means that roughly 14.5% of the U.S. population had difficulty at some time during the year providing enough food for all their members due to a lack of resources. This is uncomfortable, but not deadly. However, they also found 5.7% of U.S. households (7.0 million households) had very low food security. These people are actually already in trouble from a food system perspective despite living in nearly the richest country in the world in all of history. This means that the food intake of some household members was reduced and normal eating patterns were disrupted at times during the year due to limited resources, which are primarily economic. Perhaps most disturbingly this team from the Economic Resource Service wing of the USDA found that children were food insecure at times during the year in 10% of households with children so 3.9 million American households were unable at times during the year to provide adequate, nutritious food for their children. The historical trend is summarized in Figure 4.2. Free school lunch programs in part help fill the gap in the U.S. food supply for children. Much of the rest of the world is not as fortunate. The current food problems in the Unites States are still relatively trivial on a global scale, as we will discuss in Chapter 9.

Percent of households

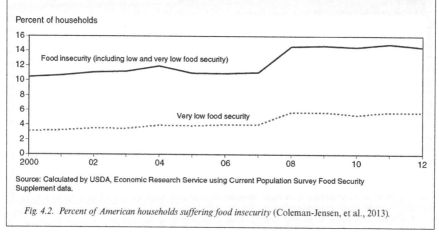

Source: Calculated by USDA, Economic Research Service using Current Population Survey Food Security Supplement data.

Fig. 4.2. Percent of American households suffering food insecurity (Coleman-Jensen, et al., 2013).

and of itself. Reducing food waste during a major crisis is one of the easiest methods of feeding more people for a longer time. However, it is not completely clear that reducing food waste now would have a net benefit during a crisis. Presumably, the economy would adjust and produce less food total. This would potentially leave humanity more vulnerable to shocks to the system because of the necessary time to ramp up production.

To remain conservative, we ignored a number of possibilities to increase the food production further:

1. Fertilizing the oceans to increase fish and seaweed production.
2. Shifting land that produces animal forage crops to human-edible crops. It should be noted that this represents a substantial gain as

meat-based food conversion is incredibly inefficient. Cattle require 6.8 pounds of feed to produce 1 pound of body mass, pigs 2.9 pounds, broiler chickens 1.7 pounds, and even, the relatively efficient fish needs 1.1 pounds of feed (Bourne, 2014).

3. Clearing large areas of forest on which to cultivate. It is feasible to cut down many trees, but it would be more difficult to remove the trees and even more difficult to remove the stumps for mechanized agriculture. Therefore, farmers would likely have to farm by hand, which would be difficult.

4. Reducing the acreage devoted to nonfood plants (such as cotton and lawns) and those plants that produce lower calories per acre (such as coffee). For example, during food shortages in the former Soviet Union, people planted yards with food.[1]

5. Basing animal production on excrement from other species (or conservatively from other families).

6. Extracting edible calories from the inedible parts of domesticated and wild plants.

Therefore, since it is feasible to ramp up food consumption by a factor of 3 quickly (note that **production** would not be scaled by nearly this much), roughly it would be technically feasible to support the Earth's population in a crisis that reduced production to 1/3 as much. In some ways, the latter scenario would be easier, requiring less fertilizer, pesticides, etc. However, a sun-obscuring scenario that reduced the net primary production potential due to lower temperatures and lower sunlight to 1/3 as much would be more difficult to accommodate. This is because farmers would have to relocate crops, and it is not clear that the crops would survive and thrive.

4.3 LIMITED CROP SUPPLY

Consider the case of a shorter post-event-growing season, where the global crop is only one-fifth as large as normal. Postharvest losses are currently ~35% (Godfray et al., 2010), and we assume that the sun-obscuring crises would cut the losses to one-third as much (more

---

[1] Lawns account for 40.5 million acres in the United States (Chameides, 2008), which is more than the area of Georgia. Thus, there is a very considerable essentially unused land area available for crop production in any of the less severe catastrophes.

extreme reduction for a more extreme crisis). Therefore, with 2100 kcal/day (Kummu et al., 2012) and 7 billion people, food demand would be approximately 1.5 Gt dry. Total current food production is ~3.3 Gt dry carbohydrate equivalent, so the limited crop would provide a 6-month supply of food.

The minimum annual global wheat storage is approximately 2 months at current wheat consumption (Do, 2010), and we assume that this applies for all grains. Given that global grain production is ~1.6 Gt/yr dry carbohydrate, and that human consumption is approximately 1.5 Gt/yr, this gives approximately 2 months of human consumption. In addition, we assume that there is a 4-month supply at crisis levels of consumption of food total in the following locations: households, stores, warehouses, wild animals, inner bark that is edible, other wild plants, and draft animals. This, coupled with the 6 months from the limited additional crop, gives a one-year supply.

4.4 MAXIMUM FOOD STORAGE

The maximum annual global wheat storage is approximately 6 months at current wheat consumption (Do, 2010). Therefore, there would be 7 months of full food from grain storage, and 4 months from other sources. So if all of the food were shared equitably, humanity would last for 1 year with the food we have now, then we would all starve to death. We should also note here that this is a snapshot in time – by 2050 the world's population is likely to increase by 35% and to feed that population the crop production will need to double as diets change (Foley, 2014). If the grain storage remains a constant number of months of consumption, there would actually be more food available per capita for a catastrophe in 2050.

4.5 FOOD SOLUTIONS FROM PREPPERS AND SURVIVALISTS

Two groups of people that have given some serious thought to food storage are survivalists and preppers. Survivalism is a movement of individuals or groups who are actively preparing for emergencies, including possible disruptions in social or political order, on scales from local to international. Survivalists often acquire emergency medical and self-defense training, stockpile food and water, prepare and practice

Fig. 4.3. A well stocked pantry photo by Robert Benner (CC).

the skills to become self-sufficient, and build structures (e.g., survival retreats or underground shelters or bunkers) that may help them survive a number of types of catastrophes. Of the two, the survivalists are generally more militaristic, perhaps best known for stockpiling weapons, while the preppers, a more moderate group, are better known for making modest stockpiles of food to prepare for short-term disasters. These stockpiles essentially are a more robustly stocked pantry, as seen in Figure 4.3. Following the financial crisis of 2007–2008, that has placed many Americans in uncomfortable economic situations there has been an explosion of prepper and survivalist literature.

The prepper and survivalist literature is an abnormal collection as it is a mixture of both high-quality government-funded texts meant for military personnel and dedicated-research volumes for woodsmen, but also many of the texts are rapidly collected self-published volumes, often displaying little analysis and in some cases offering dubious advice. In addition, unique among genres, many of the most hard-core survivalists prefer to publish anonymously or with pseudonyms. They do this out of concern that, if the disasters they focus on were to occur, the authors would make prime targets for bandits and raiders. All together though, the vast majority of the prepper and survivalist guides recommend some form of DIY food storage. They quote shelf lives generally of less than 2 years for all but bulk grains. These guides are generally written for small scale temporary disasters, e.g., an economic collapse, civil unrest, regional problems, etc. (Gee, 2013; Williams and Finazzo, 2014; Cobb, 2013, 2014a, b; Mountain, 2013; Bradley, 2012; Gregersen, 2012; Carr, 2011). Some prepper books are even directly focused on methods of storing food (Gast, 2012; Languille, 2013; Paine, 2013; Gregersen and Armstrong, 2013). They can be useful for making families better prepared to weather those types of challenges, but to deal with the prolonged mega disasters, the solutions essentially force middle-class preppers to rely for the bulk of their food from the pre-packaged grains that can last much longer. We will show below why that is not really a viable solution for most people even in America, let alone everyone on the planet.

The earlier food storage guides and those developed by Mormons like Layton (2002) tend to be more thorough and practical and some others (e.g. Pennington, 2013; Lang, 2012) outline how one could eat all of these stored foods, which are not a normal part of the diet in such bulk quantities. This is not a trivial challenge, as the food solutions we provide in the next Chapter have the same problem, and we will discuss that in more detail in Chapter 8.

The purely survivalist literature covering food concepts focuses on complete collapse of civilization without ecosystem collapse and thus primarily covers concepts involved in living off the land, eating wild plants, hunting, etc. The advice focuses on getting out of cities (where more than 80% of Americans live according the U.S. Census Bureau) and preparing to live in the wild and defend oneself. For many of the global catastrophes that are the subject of this book, these solutions are simply not viable. Even in the best-prepared survivalist situation with a well-armed retreat and a complete mastery of wilderness survivalist skills, the survivalist starves to death along with everyone else when the food supply runs out after a year or two (None of the books we surveyed recommended storing five years of food. Even the standard Vivos package which costs $20,000–$50,000/person for a luxury disaster bunker only supplies 1 year of food in their standard model (Vivos, 2014)). Thus the survivalist literature will not be summarized here. It is useful for the class of disasters for which it is focused upon. However, none of it is relevant in the mega-disasters we are focusing on in the book as the vast majority of the wildlife and plants would be dead without the sun before any of the sun killing disasters cleared. Other texts like *The Knowledge* (Dartnell, 2014) discuss rebuilding civilization after a global civilization collapse. These are academically interesting, but they would be unnecessary if the solutions we outline in the next Chapters are used to feed everyone even in a mega disaster – we can prevent such a collapse. Next let us consider the viability of using the prepper food storage solution for these extreme disasters and actually storing five years of food as individual families.

4.6 BEYOND MORMON PREPAREDNESS: PRACTICAL LIMITATIONS TO STORING 5 YEARS OF FOOD

Mormons are a religious group and the principal branch of the Latter Day Saints (LDS).[2] One of the unique aspects of the LDS Church is that it strongly encourages their member to be prepared for all types of disasters. Mormons are encouraged to store a year-long supply of food and, in a sense, it is a religious group composed of preppers. The LDS website explains: "Our Heavenly Father created this beautiful earth, with all its abundance, for our benefit and use. His purpose is to provide for our needs as we walk in faith and obedience. He has lovingly commanded us to 'prepare every needful thing' so that, should adversity come, we may care for ourselves and our neighbors and support bishops as they care for others." (LDS, 2014). The LDS Church encourages members to start with a 3-month supply and work up to a year by storing some foods that are part of the normal diet and rotating them regularly to avoid spoilage (LDS, 2014). The LDS has a relatively mature preparedness presence on the web in the Another Voice of Warning (AVOW) with more than 13,000 members, 110,000 topics and more than a million comments that generated the LDS Preparedness Manual (2012).

---

[2] Neither of us are Mormons and so this should not be taken as an endorsement for the LDS or any of their practices. On the other hand, we mean no disrespect to the LDS or its members. We are simply using them as a good example of a large group of people that take food storage seriously.

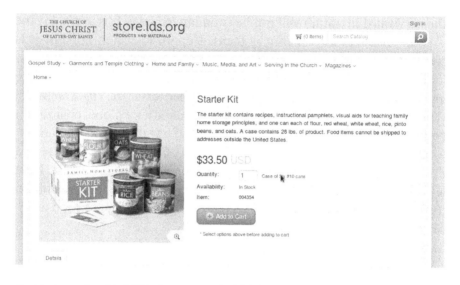

The LDS also maintains an online store where they sell pre-packaged food that can store for 20 years. For example, their starter kit contains recipes, instructional pamphlets, visual aids for teaching family home storage principles, and one package each of flour, red wheat, white wheat, rice, pinto beans, and oats. As seen in Figure 4.4 a case contains 28 lbs. of product and costs $33.50 (LDS Store, 2014), which is quite reasonable compared to what survivalist food stores are sold for on the Internet. Although the prices of bulk stored food can be reasonable, it is prohibitive for most Americans, let alone most of the world's people.[3] Costco offers a storage package that contains one hundred and twenty #10 (gallon size) cans for $1,500, which is one person-year food supply as shown in Figure 4.5. So a family with a mom, dad and two teenagers would need around $3,500 with Costco's best bulk purchase deal (Costco, 2014) and $2,898 with Sam's Club (Figure 4.6) (Sam's Club, 2014). Very few families have the additional income and financial savings to accommodate such an expense. It should be noted that although these prices are quoted from Costco and Sam's Club, bulk food for storage is still a niche market and it would appear reasonable that the prices could be reduced if a large number of people began

---

[3] If you are feeling poor, compare your annual income to the average for other countries. There are only a handful of countries where the average citizen earns what we pay for minimum wage. Billions survive with far less. Countries like India (population: 1.2 billion) have an average annual income of only $1,550.

Fig. 4.5. A screenshot of Costco's offering of a $1,500 one person-year food supply. http://www.costco.com/9687-Total-Servings-1-Person-1-Year-Food-Storage.product.100003

purchasing quantities of food for storage. Furthermore, central storage such as grain silos would be more economical. Still, it would be far more expensive than doing alternate food source research and preparation, and only implementing the alternate food sources if a disaster occurs. Furthermore, if mass food storage happened, the price of food would increase.

Fig. 4.6. A screenshot of Sam's Club Emergency Food Storage Kit that provides 1 year of food for four people.

Though this would benefit the poor who are farmers, it is likely that more people would die in the near future from starvation and hunger-related disease. As Fox News reports the median American savings is a disturbingly low $0 (Rogers, 2014). However, even if a family were to do this and manage to store 1 year's worth of food in all of the serious scenarios we discussed in Chapters 2 and 3, they would still starve to death in the second year of the catastrophe. To survive any of the disasters, the family would need to be a "Mormon family on steroids" and save enough for 5 years. Our hypothetical 4-person family would need to invest over $12,000 on stored food – enough to completely fill a bedroom. This is an unrealistic solution for 99.9% of the world's population. Interestingly enough, some of the heirs to the Wal-Mart fortune (top 0.00001% of the world in terms of wealth) have a bunker stocked up for just such an occasion.[4] Although the Waltons could easily afford to store 20 years of food for themselves, their friends and their domestic employees and Sam's Club does offer a remarkably wide selection of year-long supplies of food, it does not solve the problem for everyone. This book solves the problem of food supply for everyone else. We should point out, however, that having some stored food can help provide resilience to disruptions in the food supply chains and during lesser catastrophes (e.g., pandemics). In the final Chapter, we will discuss what individuals can do to prepare, having a small amount of stored food will be a benefit, as food storage provides more time to get started the other technical solutions that this book offers.

4.7 SURVIVALISM AND CANNIBAL MATHEMATICS

Warning: Those who are squeamish might want to skip this section without loss of continuity. All the previous scenarios assume that people are moral and share their food (although the survivalists arm themselves in the event that their storage is raided). What we did not consider is if people simply eat one another. In some past cases of starvation people have done the unthinkable and begun eating each other. For example, during the famine in Russia in 1921 some people turned to cannibalism to survive (see Figure 4.7). There are numerous historical texts of humans disturbingly turning to cannibalism, but surprisingly there is even a modern book about it (Contingency Cannibalism: Superhardcore Survivalism's

---

[4] See Robert Greenwald's 2005 documentary movie, *Wal-Mart: The High Cost of Low Price*.

Fig. 4.7. Clearly unhappy cannibals posing with their victims during Russian famine of 1921 (Public Domain).

Dirty Little Secret by Shiguro Takada, 1999). What if everyone turned to cannibalism and only the strongest would live to eat the next cycle? Could this grisly outcome help? In fact, not really as we will show below for all cases. In this non-preemptive cannibalism the population would be cut in half about every month. The population would decline rapidly and we would fall below the inbreeding limit of about 50 people in about 2 years. The end of humanity would not be pretty. We would not even make it that long if there were no vitamin C supplements.

Humanity could last longer if a large group systematically exploited cannibalism by immediately killing and preserving 5.3 billion people for food. Whoever organized such an undertaking would be responsible for more than 480 times more human death than Hitler during the Holocaust genocide in which approximately eleven million people, including six million Jews, were killed by the German military. Survival for some would be possible, although our humanity would be lost. Those that were physically alive would need to feed only on human flesh for five years. If so, 1.7 billion people could survive the full five years until sunlight was restored.

Perhaps the least morally and ethically unacceptable solution would optimize the number of people surviving without resorting to cannibalism. To do this 5.5 billion people would need to be murdered / have their food cut off and the remaining 1.5 billion people survive for the five year span. This still is not a particularly good solution, nor really acceptable from any moral or ethical standpoint, when there are better solutions as we will show in Chapter 7.

Many survivalists are hurriedly preparing to survive in such a scenario where the extreme lack of food would have starving hoards pounding on the doors of the prepared. The vast majority of survivalists are good people and only want to protect themselves and their families. Many of them even go out of their way to help their communities prepare as best they can. They look at the state of world and are trying to prepare for the worst. However, to rely on stored food and make any of the above scenarios work you would need to kill at least 3.6 people on average to survive (or you would need to kill 15 people to protect your family of 4). If most of the food gets eaten by people who will eventually starve and just a few people have protected food, each surviving person would have to kill far more people. This might be acceptable to some of the most psychotic survivalists, but they would not choose it unless it was absolutely necessary. Barring the extremely small percentage of Jeffrey Dahmers in the population, none of these solutions are acceptable.

The good news is none of them are necessary as in the next three Chapters we will show how humanity could survive even in the worst case scenarios without killing or eating anyone. In addition, everyone gets to survive – not just the weapons-hoarding cannibals and billionaires.

REFERENCES

Bourne, J.K., June 2014. How to Farm a Better Fish. National Geographic.

Bradley, A.T., 2012. The Preppers Instruction Manual. Disaster Preparer.

Brown, L.R., 2008. Why ethanol production will drive world food prices even higher in 2008. Earth Policy Institute, 24.

Carr, B., 2011. The Prepper's Pocket Guide. Ulysses Press.

Chameides, B., 2008. Stat Grok: Lawns by the Numbers. The Huffington Post. Retrieved August 6, 2014, from http://www.huffingtonpost.com/bill-chameides/stat-grok-lawns-by-the-nu_b_115079. html.

Cobb, J., 2013. The Preppers Complete Book of Disaster Readiness. Ulysses Press.

Cobb, J., 2014b. Prepper's Long-term Survival Guide. Ulysses Press.

Cobb, J., 2014a. Countdown to Preparedness. Ulysses Press.

Coleman-Jensen, Alisha, Mark, Nord, Singh, Anita, September 2013. Household Food Security in the United States in 2012. Economic Research Report No. (ERR-155), 41.

Costco. 32,000 Total Servings 4-Person 1-Year Food Storage. http://www.costco.com/32%2c000-Total-Servings-4-Person-1-Year-Food-Storage.product.11763436.html.

Dartnell, L., 2014. The Knowledge: How to Rebuild Our World from Scratch. Bodley Head.

Do, T., Anderson, K., Brorsen, B.W., 2010. The World's wheat supply. Oklahoma Cooperative Extension Service.

Foley, J.A., et al., 2011. Solutions for a cultivated planet. Nat. 478, 337–342.

Foley, J., May 2014. A Five-Step Plan to Feed the World. National Geographic.

Gast, S., 2012. Easy Food Dehydrating.

Gee, S., 2013. Economic Food Storage Strategies for Disaster Survival. Howson Books, UK.

Godfray, H.C.J., Beddington, J.R., Crute, J.R., Haddad, L., Lawrence, D., Muir, J.F., Pretty, J., Robinson, S., Thomas, S.M., Toulmin, C., 2010. Food security: The challenge of feeding 9 billion people. Sci. 327, 812–818.

Gregersen, Armstrong, 2013. Food Storage: Preserving Meat. Dairy and Eggs

Gregersen, S., 2012. Poverty Preppers.

Gunders, D., 2012. Wasted: How America is losing up to 40 percent of its food from farm to fork to landfill. Natural Resources Defense Council, http://www.nrdc.org/food/files/wasted-food-ip.pdf.

Iowa State University Extension 2006. Moisture basis conversions for grain composition data.

Kantor, L.S., Lipton, K., Manchester, A., Oliveira, V., 1997. Estimating and addressing America's food losses. Food Rev. 20, 2–12.

Kummu, M., Moel, H. de., Porkka, M., Siebert, S., Varis, O., Ward, P.J., 2012. Lost food, wasted resources: Global food supply chain losses and their impacts on freshwater, cropland, and fertiliser use. Sci. Total Environ. 438, 477–489.

Lang, A., 2012. The Prepper's Pantry:. Building and Thriving with Food Storage.

Languille, J., 2013. Prepper's Food Storage. Ulysses Press.

Lappé, F.M., 2010. Diet for a small planet. Random House LLC.

Layton, P., 2002. Emergency Food Storage & Survival Handbook. Three Rivers Press, NY.

LDS. Food Storage https://www.lds.org/topics/food-storage LDS Store. http://store.lds.org/webapp/wcs/stores/servlet/Product3_/15859395_10557_3074457345616706370_-1__195787 (visited 7/13/2014).

LDS., 2012. LDS Preparedness Manual v. 8 Book 2 Temporal Preparedness. General Membership Edition.

Meadows, D., Randers, J., Meadows, D., 2004. Limits to Growth: The 30 Year Update. White River Junction, VT, Chelsea Green Publishing Company.

Mountain, J., 2013. Hidden Harvest.

Oerke, E.C., 2006. Crop losses to pests. J. Agric. Sci. 144, 31–43.

Paine, R., 2013. Preppers Pantry. A Survival Food Guide.

Pennington, T., 2013. The Prepper's CookBook. Ulysses Press.

Rogers, K. Median American Savings: $0, Fox Business. http://www.foxbusiness.com/personal-finance/2014/05/14/median-american-savings-0/(visited 7/13/2014).

Runge, C.F., Senauer, B., 2007. How biofuels could starve the poor. Foreign Affairs, 41–53.

Sam's Club. Augason Farms Emergency Food Storage Kit (1 year - 4 people). http://www.sam-sclub.com/sams/augason-farms-emergency-food-storage-kit-1-year-4-people/prod2411209.ip?navAction=.

United States Department of Agriculture. 2013. World agricultural supply and demand estimates.

United States Department of Agriculture. 2006. USDA national nutrient database for standard reference release 17.

Vivos. 2014. Vivos. Available http://www.terravivos.com/ visited 8.15.2014.

Wallace, J., 2000. Increasing agricultural water use efficiency to meet future food production. Agric. Ecosyst Environ. 82, 105–119.

Williams, S.B., Finazzo, S., 2014. The Prepper's Workbook. Ulysses Press.

CHAPTER 5

Stopgap Food Production: Fast food

5.1 THE 10 °C CRISIS AND THE 20 °C CRISIS

Many of the global disasters discussed in Chapters 2 and 3 would reduce agricultural productivity less than ~70%; therefore, the technical solution for feeding the entire world's population would be to reduce the amount of preharvest losses, the yield underachievement, wasted food, edible food fed to livestock and pets, and edible food turned into biofuels, as we discussed in Chapter 4. As the severity of a crisis increases, humanity would require more extreme measures to prevent widespread starvation. For example, in a full-scale nuclear winter simulation (Robock et al., 2007), the maximum global temperature reduction was 9 °C, which we will round up and call the "10 °C crisis." Farmers could grow a considerable amount of food after the crisis first starts, as there is a lag before temperature and ozone are fully reduced (Mills et al., 2008). This, coupled with food storage, would provide full human food [1.5 billion tons (Gt)/yr] for the first year and allow other sources to be ramped up ("slow food"). We provided an example of such a situation in Chapter 4.3.

However, if nearly all the sunlight is blocked, and therefore, the temperature drop would be roughly double to what we will call "the 20 °C crisis," the additional crop would be negligible. If the crisis hit at maximum food storage, the storage would be approximately 11 months of food. The maximum annual global wheat storage is approximately 6 months at current wheat consumption (Davis et al., 2009), and we use the assumptions of Chapter 4.3. Therefore, there would be 7 months of full food from grain storage, and 4 months from other sources. The worst-case scenario of the 20 °C crisis hitting at minimum food storage time with little warning would require unconventional food in less than 1 year. We call this "fast food." We focus on feeding the entire global population for a 5-year time horizon because some of the less promising solutions are likely to produce an important amount of food after 5 years, and some crises last less than 5 years.

Feeding Everyone No Matter What. http://dx.doi.org/10.1016/B978-0-12-804447-6.00005-X

5.2 STOPGAP FOOD PRODUCTION: FAST FOOD

Some of the unconventional food supplies in Chapter 6 take about 1 year to ramp up and the minimum stored food will only last ~6 months. Therefore, for the 20 °C crises, there is a need for stopgap food production of "fast food." There are three main solutions, all utilizing nonwoody and less than ~1 cm thick woody ("thin") biomass for fast processing: (1) extracting edible calories, (2) mushrooms, and (3) bacteria.

The human-digestible fraction of the dry weight of killed tree leaves, examples of which are shown in Figure 5.1, (as opposed to depleted leaves that are shed as leaf litter) is approximately 50% (Jacquemoud et al., 1996) and we assume this applies to all nonwoody biomass that is killed. Because of all the fiber, the net calories would be relatively small because the digestion of fiber decreases the energy yield from eating food. Thus, some caution is needed for using high-fiber-material-conversion efficiencies. However, three ways around this problem are (1) making tea, (2) chewing (but not swallowing the solids), and (3) grinding and pressing leaves, and then, coagulating (causing the solids to clump for removal) the resultant liquid.

Fig. 5.1. Killed leaves for dinner. Photo credit: E. Bow Pearce (Public Domain).

5.2.1 Exacting Food from Leaves

For black tea, about 1/5 of the calories in the sum of the protein, carbohydrate, and lipid (fats and oils) make it into the liquid (Belitz et al., 2009). If this were true for all nonwoody biomass, it would correspond to 10% conversion efficiency and people already drink pine needle tea (Kim and Chung, 2000). An alternative technique is grinding and pressing leaves, and then coagulating the resultant liquid. The resultant leaf concentrate contains ~8% of the dry matter of the original leaves with the industrial process (Leaf for Life, 2013). The remaining liquid contains much of the toxins and is unfit for human consumption. However, industrial

Information Box 5.1

Pine needle tea, as shown in Figure 5.2, has 400%–500% of the vitamin C of fresh squeezed orange juice and is high in vitamin A. Thus, it would be extremely valuable in scenarios that would otherwise bring on scurvy (e.g., consequences include your teeth falling out like old pirates). To make the tea, gather 1 part pine needles to 3 parts water. Rinse and chop the pine needles. Then, either add needles to boiling water and let it simmer for 10 min or add pine needles to bottom of cup, pour in boiling water, and steep for 20 min. Strain liquid into cups and (if you have it) add a sweetener.

Fig. 5.2. Fresh pine needle tea. Photo credit: J. Bow Pearce (Public Domain).

techniques may be able to reclaim additional edible calories from this liquid. The yield of the leaf concentrate is lower with nonindustrial techniques, such as automobile jacks for pressing (Kennedy, 1993). Current techniques utilize fresh-cut leaves; some dried and rehydrated leaves may have a bitterness problem. Therefore, we estimate 5% of calorie extraction overall.

This tea/pressing technique could be used on inner bark or even on ground sapwood (the living part of wood, as opposed to the dead heartwood). This is because sapwood in killed trees contains cells that were recently alive and some beetles that cannot digest cellulose (the three major parts of fiber or lignocellulose are lignin, cellulose, and hemicellulose) can still get their nutrition from sapwood (see Figure 5.3). Sapwood would likely only be a partial solution because of limited grinding capability, but then enzymes or other food production methods

Fig. 5.3. A section of a yew branch showing the sapwood and dark heartwood and pith (center dark spot). The dark radial lines are small knots. Wikimedia commons (CC).

could utilize the residue. It may be possible to press cooked branches even without chipping or grinding.

5.2.2 Supply: Global Nonwoody Vegetation

We quantified the vegetation carbon mass and multiplied it by two to get dry biomass (Hoekman et al., 2010). We assume that the nonwoody biomass was made up of temperate grasslands, deserts and semi-deserts, tundra, wetlands, and croplands; there would be some trees in these habitats, but there would also be nonwoody biomass in the other habitats of tropical forests, temperate forests, boreal forests, and tropical savannas (IPCC, 2000).

There may also be branches from the trees people cut down for the "slow food" production, woody shrubs, and short trees as possible sources of thin biomass. However, drying of the wood would take considerably longer, so given the short time horizon, the weight of the material to ship would be considerably greater than the dry biomass in the leaves that can be converted quickly. Removing the leaves by hand would likely be too labor intensive, but industry could retrofit light-duty vehicle (cars, SUVs, and pickup trucks) manufacturing capability to produce machines to achieve this task. There is also considerable biomass in the form of leaf litter. The leaves and needles are much smaller than the grass that existing equipment can harvest, but retrofits or new equipment could address this issue. Also, another option is relocating the ruminants to the areas with leaf litter. Ruminants are mammals that obtain nutrients from plant-based food by fermenting it in a specialized stomach prior to digestion, principally through bacterial actions. The process typically requires regurgitation of fermented ingesta (known as cud), and chewing it again. Ruminants include common animals like cows, goats, sheep, yaks, deer, camels, and llamas. The difficulty with ruminants is that they cannot scale up fast enough to be a full fast food solution, as we will show in Chapter 6. Therefore, we conservatively ignore these partly woody and leaf litter sources of thin biomass for the immediate fast food, but these could facilitate thin biomass as a full solution.

In total, the global nonwoody vegetation above ground is ~90 Gt; thus, even if only 1/3 of this is mechanically harvestable and transported to population centers quickly, this is 30 Gt. If half is nontoxic with 5% dry matter extraction, it would be a half year supply of food. Industrial processing could counteract inedibility of certain plants. Additionally, the remaining material could be fed to bacteria or mushrooms, extending the supply. This could maintain food supply despite a lower amount of biomass harvest, such as would occur if the disaster struck during the northern hemisphere winter.

5.3 MUSHROOM FAST FOOD

For mushroom fast food, approximately 1/3 of existing building space would be required to grow all human food, though there would be complications. Following the assumptions on mushroom growth we will go into detail on in Chapter 6, white button mushrooms typically yield 20 kg/m^2 (Kaul and Dhar, 2007) so we assume half of this. The typical yield time is 3 months, so we assume 50% longer. The high relative humidity in buildings caused by trays growing bacteria or mushrooms is an important problem because of mold growth. This can be mitigated by sterilizing and sealing susceptible surfaces. Also, using dedicated buildings for food production could limit human mold exposure. Feasible examples of this include office buildings where the people switch to telecommuting and industrial buildings that are no longer operating for their original purpose. The growing of bacteria or mushrooms indoors would also cause high carbon dioxide levels. Dedicated buildings would also limit human exposure. If these buildings are insufficient, basements could be isolated more completely from the rest of the housing units. Reduced oxygen levels are less of an issue because of the small relative change. This would generally allow normal activities to continue and is conservative because it is ignoring the possibility of using new or temporary structures, caves, and existing mines. Conservatively assuming no human weight loss, the mushrooms have to provide about 5 months of food. White button mushrooms have a maximum biological efficiency (wet weight of mushrooms divided by dry weight of feedstock) of 100% (Chang and Miles, 1984) so with 90% water, this is ~10% caloric efficiency. We assume half the 10% value due to nonideal substrates and pests. The conversion efficiency would then be ~5%. Figure 5.4 summarizes the percent of human food as a function of time for both stored food and the mushroom solution.

As can be seen in Figure 5.4 the stored edible food at first provides 100% of human food needs but then quickly is reduced to under 0.1% in

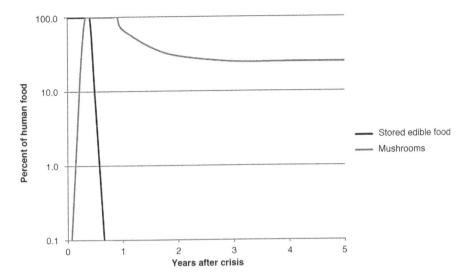

Fig. 5.4. Percent of food for all of humanity over time for stored food and the mushroom solution. Note that the logarithmic vertical scale, where each horizontal line is a factor of 10 greater. Note also that slow mushrooms (outdoor on trees) contribute food after 1 year.

less than a year. Conversely, mushroom production cannot meet all of humanity's food requirements now (zero on the X-scale of the graph). However, the mushroom solution can quickly ramp up to provide all of human food needs for a time before being reduced to do growth supplies. This is a fairly monotonous diet, but would provide enough calories to keep humanity going until the slow food of Chapter 6 is ready. But, what if you are allergic to mushrooms or like one of our wives – just does not like mushrooms? We have both you and her covered.

5.4 NOT QUITE AS GOOD AS MUSHROOMS – BACTERIA TO HUMANS FAST FOOD

The less palatable fast food solution is humans eating mostly-bacteria-digested thin biomass (see Figure 5.5), which tastes as good as it sounds. The good news is, however, with the rapid doubling of bacteria in a favorable environment, this solution could provide 100% of food after only 2 months. The fiber should be relatively low because the bacteria can be mixed throughout Nutrition from bacteria could be an issue and needs to be explored further (as discussed in Chapter 10).

There may be a solution that increases conversion efficiency and palatability. The process would include chipping or grinding wood and

Fig. 5.5. Partially digested plant material. *Original picture compliments of the DOE Great Lakes Bioenergy Science Center under the U.S. Department of Energy Genomic Science program* (Public Domain).

having bacteria growing that secrete cellulase, which turns the cellulose into sugar outside their bodies. But instead of waiting until the bacteria absorbs the sugar, running water through the mixture could leach the sugar out. If humans only consumed the sugar water, the mushrooms (which can digest the lignin) could grow on the leftover material, and ruminants could eat the leftover from that. The industrial ecology scheme is illustrated in Figure 5.6 and the summary of the fast food source as a function of time is shown in Figure 5.7. It should be noted that extracting food from leaves rises the fastest and then falls quickly (shown in Figure 5.7). Following this model, at least some of the humans would continue to survive off of fairly attractive foodstuffs (e.g., steak and

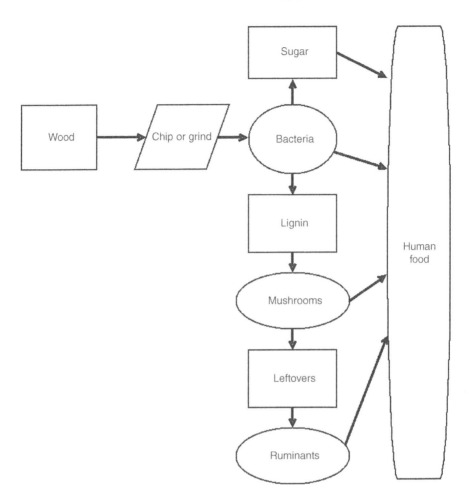

Fig. 5.6. Industrial ecology scheme for fast food including (1) chipping or grinding wood, (2) bacteria growing on the wood that secrete cellulase, which turns the cellulose into sugar outside their bodies, (3) running water through the mixture leaches the sugar out to feed humans, then (4) mushrooms digest the lignin, which humans then consume, and (5) the leftover material is fed to ruminants that are then eaten by humans as well.

pop/soda by carbonating the sugar water) in the near term. If humans consume the bacteria and fiber, this would balance the macronutrients better. Finally, another possibility is industrially separating the bacteria from the fiber.

It is possible that these thin biomass sources could become a full solution on their own. The longer time would allow the harvest and transportation of considerably more biomass. Multiple organisms could consume wastes in series from the same biomass (including rats and chickens as they ramp up) as we will discuss in detail in Chapter 6.

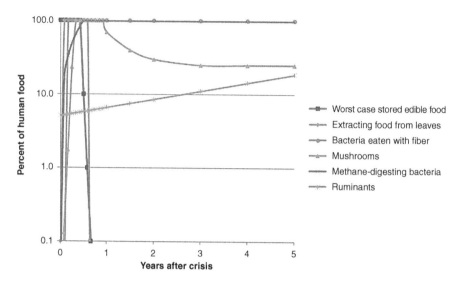

Fig. 5.7. Ramp rate summary (percent of human food) of fast food solutions as a function of time. It should be noted that extracting food from leaves and bacteria eaten with fiber rise so quickly that the markers every month for the first year are not visible. Also, methane-digesting bacteria are discussed in Chapter 7.

REFERENCES

Belitz, H.D., Grosch, W., P., Schieberle, 2009. Food Chemistry. Springer, Berlin.

Chang, S.T., P.G., Miles, 1984. A new look at cultivated mushrooms. Bioscience 34, 358–362.

Davis, S.C., Anderson-Teixeira, K.J., E.H., DeLucia, 2009. Life-cycle analysis and the ecology of biofuels. Trends Plant Sci. 14, 140–146.

Hoekman, S.K., Broch, A., Robbins, C., Zielinska, B., Coronella, C.J., L.G., Felix, 2010. Energy densification via hydrothermal pre-treatment of cellulosic biomass. AWMA International Specialty Conference: Leapfrogging Opportunities for Air Quality Improvement.

Intergovernmental Panel on Climate Change (IPCC), 2000. Land Use, Land Use Change and Forestry. Cambridge University Press, United States.

Jacquemoud, S., Ustin, S.L., Verdebout, J., Schmuck, G., Andreoli, G., B., Hosgood, 1996. Estimating leaf biochemistry using the PROSPECT leaf optical properties model. Remote Sens. Environ 56, 194–202.

Kaul, T.N., B.L., Dhar, 2007. The Biology and Cultivation of Edible Mushrooms. Pennsylvania State University, Westville.

Kennedy, D., 1993. Leaf concentrate: A field guide for small-scale programs. Leaf for Life. http://www.leafforlife.org/PDFS/english/Leafconm.pdf

Kim, K.-Y., H.-J., Chung, 2000. Flavor compounds of pine sprout tea and pine needle tea. J. Agric. Food Chem. 48, 1269–1272.

Leaf for Life, 2013. Industrial leaf concentrate process (France). Available: http://www.leafforlife.org/pages/industri.htm.

Mills, M.J., Toon, O.B., Turco, R.P., Kinnison, D.E., Garcia, R.R., 2008. Massive global ozone loss predicted following regional nuclear conflict. Proc. Natl. Acad. Sci. USA 105, 5307–5312.

Robock, A., Oman, L., G.L., Stenchikov, 2007. Nuclear winter revisited with a modern climate model and current nuclear arsenals: Still catastrophic consequences. J. Geophys. Res. Atmos. 112, 1984–2012.

CHAPTER 6

Fiber Supply for Conversion to Food

6.1 FIBER SUPPLY FOR CONVERSION TO FOOD

If we were to attempt to feed humanity by converting existing (and soon to be dead) biomass into food following a global catastrophe, how much do we have to work with? To answer that question, first we find that the stock of dry biomass in vegetation is approximately 1200 Gt (Food and Agriculture Organization, 2000). As the global tree harvest (e.g., round-wood, as shown in Figure 6.1) is 3 billion m^3 (O'Driscoll, 2008) with a dry density of 0.5 tons/m^3 (Ragland et al., 1991), that makes available 1.5 Gt/yr at current harvesting rates. However, if the planet cooled considerably, a lot of it would freeze at least part of the year. In the 10 °C crisis case, roughly half of this vegetation would be in areas that did not freeze. In the 10 °C crisis simulation, nearly all of the tropical forests had an average temperature loss of less than 20 °C (Robock et al., 2007). In these tropical areas, the mean temperature would be around 5 °C, and with the roughly half light levels, the annual and diurnal variations would be relatively small. Thus, since 57% of the global forests are tropical (IPCC, 2000), and since nearly all vegetation is trees, roughly half of the vegetation would be in areas that did not freeze. The peak temperature response is roughly proportional to the peak reduction in shortwave (solar) radiation for both the sulfate (Timmreck et al., 2012) and soot (Robock et al., 2007) crises. Therefore, in the 20 °C crisis, roughly half of the tropical forests would remain above freezing (Robock et al., 2007). This represents an enormous amount of biomass to convert to food, but whether that is sufficient to feed everyone for 5 years depends on the conversion efficiency.

Although more difficult, it is possible that humans could utilize below-ground biomass. One option is using burrowing animals like beetles. Another option that would only apply to trees is pulling trees down, because this often uproots the tree (as shown in Figure 6.2) rather than cutting it down. However, we conservatively ignore these options.

Feeding Everyone No Matter What. http://dx.doi.org/10.1016/B978-0-12-804447-6.00006-1

Fig. 6.1. Stacked logs waiting to be transported out of the forest. Photograph provided by Patrick Mackie (CC).

Fig. 6.2. Uprooted tree over the trail at Elizabeth A. Morton National Wildlife Refuge, part of the Long Island National Wildlife Refuge Complex. Photo credit: Todd Weston/USFWS (CC).

Though there is sufficient building space to provide all of human food with thin biomass (see last Chapter), the conversion rate of trees is much slower. Humans could build temporary structures, but this would require stacking of the logs to utilize this volume better. There is currently limited stacking capacity, so we leave ramping calculations to future work. Therefore, to remain conservative, the only log scenario we consider is tree conversion outdoors in areas where it does not freeze.

To get at this wood fiber, we need to be able to cut down the trees (they would die anyway so environmentalists need not be alarmed) and so we need a lot of chainsaws (e.g., see Figure 6.3). Global sales of chainsaws are around 3 million annually (Koehler et al., 2004). We assume that chainsaws have a lifetime of 10 years, giving a stock of 30 million chainsaws. Production rate for Juniper is approximately 2.5 dry tons per hour (McNeel and Swan, 1997). Juniper is more difficult than the typical tree, but in most cases, loggers would have to use residential chainsaws instead of commercial chainsaws. Therefore, we assume the typical tree would also be 2.5 dry tons per hour. Additionally, we assume 80% duty cycle on chainsaws. Utilization of current chainsaws could thus

Fig. 6.3. A large and varied collection of chain saws being displayed at the Wisbech St. Mary country show. Photo credit: Michael Trolove (CC-BY-SA).

fell approximately 500 Gt dry of trees/yr. However, given that ~40% of the total mass of wood is above ground, the chainsaw total is considerably greater than the tropical forests [~170 Gt above ground (Food and Agriculture Organization, 2000)].

Since we show in Chapter 7 that there is sufficient methane and accessible above-ground wood for the food supply for 5 years, humanity would not have to utilize less desirable fiber sources. These include below-ground wood, soil carbon (including peat), landfills, houses, and hydrothermal vents. We include them here just for completeness.

6.2 WORST CASE: 20 °C CRISIS FIBER AVAILABILITY

Let us first consider the 20 °C crisis. Freight capacity could transport logs from unfrozen zones, but this would only facilitate outdoor mushrooms, which is not a complete solution (see Chapter 7). Therefore, humanity would probably use the transportation capability for other food sources instead. The case of losing 20 °C would result in approximately 80 Gt available above ground that did not freeze. The current global shipping traffic is 53,000 Gt-km (UNCTAD, 2009). The current U.S. freight transportation is 9,000 Gt-km, 41% of which is trucking (Office of Freight Management and Operations, 2012). If global capacity is 400% more, the global trucking capacity is 19,000 Gt-km. However, the duty cycle of trucks is considerably lower than ships, ships' duty cycle can be increased somewhat, and there are other (less flexible) modes, so we estimate a total capacity of 100,000 Gt-km.

Because humanity would generally dry the food, the critical transportation requirement is the feedstocks because their mass is generally much greater than the food produced. The natural gas transportation capability would already be adequate for the methane-digesting bacteria detailed in Chapter 7. It would not be prudent to spend transportation capability transporting logs for the 20 °C crisis. For rat, chicken and beetles solution for the 20 °C crisis that will be discussed in the next Chapter, humanity must transport chipped wood to buildings. The conversion efficiency is around 5%, so over 4 years, this requires only one-third of the above-ground trees. Therefore, the transportation distance would be moderate. However, in the case of thin biomass providing a full solution, the humanity would have to harvest biomass from the entire world, so the transportation distance would be high. Therefore, this is the critical case. Since the Americas have lower population density, there is a regional mismatch of humans and biomass. Therefore, the transportation distance could grow to 15,000 km. With roughly 10% efficiency associated with indoor mushroom conversion followed by another animal step, ~15 Gt/yr would be required. The transportation requirement would then be double current capacity at high duty cycle. However, with retrofitting automobile manufacturing facilities for freight transportation equipment, industry could meet this requirement.

6.3 WOOD CHIPPING

Although there are residential chippers, road chippers, and chippers for paper and paperboard production, retrofitting existing manufacturing facilities to produce wood chippers could increase substantially the chipping capacity. The current global production rate of light-duty vehicles is 63 million/yr (International Organization of Motor Vehicle Manufacturers, 2013). The typical engine peak output is ~100 kW mechanical (Ma et al., 2012), and we assume one-third of this for continuous operation (with additional forced air cooling). In World War II, the U.S. retrofitted 66% of automobile manufacturing for airplane production in less than 2 years (Zeitlin, 1995). Since the crisis considered here would be more serious, we assume that industry would retrofit nearly all automobile manufacturing facilities over the course of 1 year. Then, by the end of the year, this would produce a wood chipping capacity of ~1 TW mechanical. With 80% duty cycle, and 180 J/g to chip wood (Zhu and Pan, 2010), this is ~100 Gt of wood chipped in the first year. This is sufficient to provide full human food for all solutions discussed in the next Chapter except for outdoor mushrooms. Alternatively, current light-duty manufacturing could go primarily to producing wood grinders to produce smaller particles for enzymatic conversion. Under the same scenario above except for 1980 J/g to grind rather than chip wood (Zhu and Pan, 2010), this would be ~10 Gt of wood ground in the first year. This would be more than sufficient to provide full human food with enzymatic conversion that has efficiencies near 30%, as we show in the next Chapter. These chipping and grinding scenarios are conservative because there is the possibility of retrofitting other manufacturing facilities to produce wood-comminution (size reduction) equipment, and there is also the possibility of retrofitting existing motors and engines into chippers/grinders. In addition, it is less energy intensive to chip partially decomposed wood, of which there would be a considerable supply initially, and a continuing supply as the trees killed by the disaster decompose.

Furthermore, it may be possible to use thermal methods for comminution. One method is steam explosion, which is currently used to break wood chips into smaller pieces and pretreat for enzymatic cellulose biofuel production. The idea is heating wet wood under pressure, and then suddenly releasing the pressure, which causes the water to boil and expand rapidly, exploding cells. This may achieve chipping as well, but would require pressure vessels. More speculative methods include raising the temperature of saturated wood quickly with hot oil or air, causing explosion.

Overall, even without any form of speculative chipping, chipping all the fiber would take a small percentage of humanity's current energy use, and it should be feasible to retrofit light-duty vehicle factories to produce wood chippers that could chip sufficient wood within 1 year.

6.4 FIRE SUPPRESSION

Once the trees die and dry out, as shown in Figure 6.4, they would be more susceptible to fire, which would require a considerable, but feasible, ramping up of current fire suppression capabilities. A much greater

Fig. 6.4. Altered photograph of forest dieback in the Bavaria Forest, which would be similar to the forest death after one of the global catastrophes discussed in previous Chapters. Photograph provided by High Contrast (CC).

Fig. 6.5. Air National Guard C-130 Hercules drops fire retardant on wildfires in southern California. In a long emergency, other private and commercial aircraft could be retrofitted for fire suppression. Photo credit U.S. Air Force, photo by Staff Sgt. Daryl McKamey (Public Domain).

area than is currently susceptible to burning would become susceptible to burning. Even though some biomass would not be viable for food production, if a considerable amount burned, this could cause additional cooling (Robock, 1991). Industry could re-purpose ground and air vehicles (e.g., see Figure 6.5) to fire fighting and produce additional vehicles by repurposing other manufacturing capability. Ramping up the production of chemicals for fire suppression shown in Figure 6.5 may be more difficult, and it could cause issues with the edibility of the food produced from the wood. So less advanced techniques could be used, like water. Therefore, we find it safe to assume no important reduction in available wood for conversion to food.

REFERENCES

Food and Agriculture Organization, 2000. Global Forest Resources Assessment. Food and Agriculture Organization, United Nations.

Intergovernmental Panel on Climate Change (IPCC), 2000. Land Use, Land Use Change and Forestry. United States, Cambridge University Press.

International Organization of Motor Vehicle Manufacturers, 2013. World Motor Vehicle Production.

Koehler, S.A., Luckasevic, T.M., Rozin, L., Shakir, A., Ladham, S., Omalu, B., Dominick, J., Wecht, C.H., 2004. Death by chainsaw: Fatal kickback injuries to the neck. J. Foren. Sci. 49, 345–350.

Ma, C., Kang, J., Choi, W., Song, M., Ji, J., Kim, H., 2012. A comparative study on the power characteristics and control strategies for plug-in hybrid electric vehicles. Int. J. Automot. Technol. 13, 505–516.

McNeel, J.F., Swan, L., 1997. Harvesting Western Juniper (*Juniperus occidentalis*) in Eastern Oregon – A case study. http://juniper.oregonstate.edu/bibliography/documents/phpItcJeh_mcneel.pdf

O'Driscoll, E., 2008. Roundwood supply, wood energy and related issues in the UNECE region. Coford Connects Processing / Products, 14.

Office of Freight Management and Operations, 2012. Freight analysis framework, version 3.4.

Ragland, K.W., Aerts, D.J., Baker, A.J., 1991. Properties of wood for combustion analysis. Biores. Technol. 37, 161–168.

Robock, A., 1991. Surface cooling due to forest fire smoke. J. Geophys. Res. 96 (20), 869–920.

Robock, A., Oman, L., Stenchikov, G.L., 2007. Nuclear winter revisited with a modern climate model and current nuclear arsenals: Still catastrophic consequences. J. Geophys. Res. Atmos. 112, 1984–2012.

Timmreck, C., Graf, H.-F., Zanchettin, D., Hagemann, S., Kleinen, T., Krüger, K., 2012. Climate response to the Toba super-eruption: Regional changes. Quat. Int. 258, 30–44.

UNCTAD Secretariat, 2009. Review of Maritime Transport 2009.

Zeitlin, J., 1995. Flexibility and mass production at war: Aircraft manufacture in Britain, the United States, and Germany, 1939–1945, Technol. Culture 36, 46–79.

Zhu, J.Y., Pan, X.J., 2010. Woody biomass pretreatment for cellulosic ethanol production: Technology and energy consumption evaluation. Biores. Technol. 101, 4992–5002.

CHAPTER 7

Solutions: Stored Biomass/Fossil Fuel Conversion to Food

7.1 SOLUTIONS INTRODUCTION

Increased food storage was the solution previously proposed to all the crises we discussed in Chapters 2 and 3 for feeding all people (Bostrom and Cirkovic, 2008). As we have shown in Chapter 4, storage is not viable in the near term and would exacerbate existing levels of mortality due to inadequate global access to affordable food – just as policies did that supported biofuels (Runge and Senauer, 2007). Due to malnutrition and hunger-related disease, 6.5 million deaths/yr occur in children under 5 years old (United Nations Children's Fund, 2006). Note that the technical solution to feeding people now is much less challenging than feeding people during a crisis – we make plenty of food (Fraser, 2014). This shows that the economics and politics of feeding people in a crisis is important future work.

For the purposes of determining all scenarios, we assume that the entire current global population, biomass (by dry weight), and infrastructure are intact. Some disasters would kill people, burn or radioactively contaminate regional biomass, and/or destroy infrastructure, but the severity of the problem would be similar. For example, if a large city were destroyed by a major asteroid impact, we would not have access to the city's infrastructure to use for feeding everyone in the resultant disaster. However, the challenge of feeding humanity would be decreased by the city's population. There are several considerations that would also shrink the scope of the problem. First, for the crop-killing scenarios, if only certain species or geographic regions are affected, other crops or areas could be substituted [and food may be extracted from the super weed or pests (see Chapter 5)]. Even in some sun-obscuring scenarios, growing some food with remaining natural light is possible. The amount of food produced from natural light would be dependent on the type and intensity of disaster and is a function of time. For a global maximum

Feeding Everyone No Matter What. http://dx.doi.org/10.1016/B978-0-12-804447-6.00007-3

10°C temperature drop (~50% light blocked), there would still be considerable areas in the tropics with a long growing season. Soot crises
that reduce the ozone dramatically would require cold and UV tolerant
plants, which do exist at high altitude, but currently provide a negligible
amount of food. Furthermore, there would likely be difficulties with
changes in humidity, photoperiod (daylight length), soil pH (e.g., acid-
ity), and soil particle size. It may be feasible to ramp up over 5 years, or
10 years even for a 20°C black carbon crisis, but this is more speculative,
so we do count on it.

7.2 SUSHI FOR DINNER?

Currently ~0.1% of the ocean area undergoes coastal upwelling bring-
ing nutrients to the surface (as seen in Figure 7.1), but it produces ~50%
of the global fish catch because of the high nutrient levels (Wallace and
Hobbs, 1977). A similar effect would occur when a sun-obscuring disaster cools the upper layers of the ocean and they sink. Therefore, since
the ecosystem would be shielded from the UV, the marine caloric contribution could be more than a year's worth of food.

The current global fish catch is approximately 0.11 Gt/yr wet [including
the part currently discarded (Pauly, 1996), which would be retained in
a crisis]. Because the finfish (e.g., tuna) and shellfish (e.g., scallops) are
~1.5 kcal/g wet (Hartman and Brandt, 1995), this is 3% of the calorie

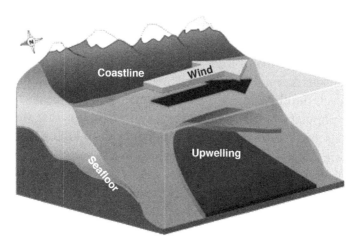

Fig. 7.1. Upwelling brings nutrients to the surface (Public Domain).

Fig. 7.2. Summit caldera of Pinatubo early eruption 1991 (Public Domain).

requirement of humans assuming eating the entire fish (this tonnage estimate includes aquaculture and freshwater fishing, but excludes unreported artisan fishing). After the Mount Pinatubo eruption in 1991 (Figure 7.2), despite lower temperatures and precipitation, and 35 W/m^2 total insolation (solar intensity) reduction [~1/6 of total 200 W/m^2 (Pearce, 2008)], global net primary productivity actually increased (Mercado et al., 2009). This is due to plants utilizing diffuse light more efficiently than direct sunlight. The 10°C crisis would have a considerably greater insolation reduction (~50%). Also, the fish production is a strong function of the plant net primary production because if there are dense plants, it requires only one trophic (food chain) level to go to fish, while in the open ocean it requires many trophic levels (Pinet, 1992). Furthermore, for the soot crises that destroy substantial amounts of ozone, there will be considerably more UV reaching the surface of the earth. In ocean water, UV is absorbed more strongly than photosynthetically active radiation (Kuwahara et al., 2000), meaning that plants would be protected from increases in UV. However, the algae growth would need to be further down in the water, where there is less photosynthetically active radiation. Therefore, we assume that the fish production per area of the entire ocean for 2 years (~1.5 years of cooling and therefore upwelling, and fish production for another half year) is one-third as much as current coastal upwelling. This would produce about 9 years of food.

A limiting factor is humanity's ability to capture the fish. The duty cycle (fraction of the hours of the year of operation) of large fishing boats is ~74% (Corbett and Koehler, 2003). We assume that only 50% of the duty cycle is actually fishing, while the remainder is transit. The fish would be much further from shore on average than currently, but ship-to-ship cargo transfer (which ships currently perform) could move the fish to freight ships in remote areas to minimize fishing-boat travel distance. Therefore, we estimate that 80% of the duty cycle would actually be fishing in a crisis.

Boats catch the vast majority of fish caught outside of the coastal upwelling regions over the continental shelf (Nellemann et al., 2008). The continental shelf makes up ~10% of the ocean area (Nellemann et al., 2008), and we estimate that based on the overlap of fish landings and the continental shelf that boats fish 80% of this shelf intensively. This yields a fish capture density of coastal upwelling of approximately a factor of 80 greater than that on the general continental shelf. If the ocean undergoes upwelling because of a global temperature loss of 10°C, the nutrient concentrations will be denser than the typical area that humans fish from currently. Therefore, the fish density will be higher, so the same boats could catch more fish. We assume that the fish caught per boat is proportional to the square root of the fish density because boats compensate for the lower fish density. This means that if the fish density everywhere were equal to one-third that of coastal upwelling, the current fishing boat fleet at current duty cycle would be able to catch ~200% more fish. Therefore, with increased duty cycle, the total fish catch would be ~20% of the total human food requirement. This analysis is conservative because industry would produce more fishing boats and retrofit other ships (including military) for fishing, given that only 4% of total ship power is in fishing boats (Corbett and Koehler, 2003). Therefore, it appears likely that full human food supply is feasible.

Figure 7.3 shows anchovies, which mature in 1 year and have approximately 50,000 eggs in the first year after maturity (Smith, 1985); therefore, if the actual growth rate is the square root of the ideal, this would be a factor ~70/yr. Since the current anchovy catch is 0.0106 Gt/yr (Eurofish, 2012), they could produce full human food in 1.5 years, which conservatively ignores marine mammals, krill (similar to shrimp), other fish, etc.

Fig. 7.3. Anchovies. Photo credit Paul Asman and Jill Lenoble (CC).

With an annual shipping capacity (with high duty cycle) of approximately 60,000 Gt-km/yr, and approximately 4 Gt of wet fish required for all human food, this could support an average transportation distance of 15,000 km. This would be more than sufficient because boats would preferentially take fish from waters near to people (the actual fish production from upwelling is greater than humans would have capacity to catch).

The oceans could produce considerably more calories via seaweed, such as the salad shown in Figure 7.4, instead of fish because seaweed converts the nutrients more fully and the percent protein is generally lower (requiring less nitrogen per calorie). However, human production of seaweed would have to be ramped up much faster than fishing. In addition, harvest would be more difficult. Furthermore, seaweed is considerably higher percent water, making it less feasible to transport long distances, as there is likely insufficient volume on a ship for drying equipment and energy. Therefore, here we very conservatively ignore the contribution of the food supply from seaweed, although it should clearly be looked at in more detail for providing increased food supplies in non crisis situations.

After the cooling upwelling stops (and for all time in the crop-killing scenarios) macronutrient fertilization of the ocean could provide full

Fig. 7.4. Seaweed salad. Photo courtesy of Vegan Feast Catering (CC).

human food. The macronutrient fertilization scenario is similar to the global cooling-induced upwelling. Fertilization would be required to maintain fish production after upwelling ceases. Furthermore, fertilization would be required from the start for the crises that do not cool the ocean. With better control of the nutrients than cooling upwelling, we estimate that fish production would be as great as current coastal upwelling, despite lower insolation and possibly higher UV. Therefore, fertilizing only a small fraction of the ocean area would produce full human food. Boats would fertilize this area preferentially close to people, so this would require even a smaller fraction of shipping capability than cooling upwelling. Therefore, the current fishing boat fleet could harvest ~30% of full human food. Full human food would be even easier in this scenario. In general, it would be more feasible to relocate an ecosystem to a lower latitude in the ocean than the land because the ocean water is more uniform than soils. However, for the more extreme disasters in which 90% or more of the light is blocked, temperatures would be even lower and photosynthesis would be severely limited, so another solution is required, and here, we investigate the solutions utilizing stored biomass and fossil fuels.

7.3 OIL AND GAS FOR DINNER? THE CASE FOR INDUSTRIAL FOOD

Synthetic food production refers to using chemical synthesis. Sugar has been synthesized from noncarbohydrates for decades (Hudlicky et al., 1996). Global fossil-fuel emissions are approximately 6 Gt carbon/yr (Pacala and Socolow, 2004). The equivalent dry carbohydrate by carbon content is 12 Gt/yr. This is roughly an order of magnitude larger on an energy basis than the necessary production of food for all humans ($\sim$1.5 Gt/yr). One option would be converting crude oil or its distillates (e.g. gasoline) into edible oil. Another option is converting natural gas to foods such as sugar. A further option involves gasifying solid fuels and then converting to food.

A hybrid industrial/biological technique would be providing chemicals industrially to chemo-synthesizing bacteria, those that oxidize methane or an inorganic chemical to produce energy. Even if it were possible to use all artificial and natural sources of hydrogen sulfide for bacteria to oxidize (with carbon dioxide as the carbon source), this is only $\sim$0.001 Gt H_2S/yr (Bates et al., 1992; World Health Organization, 2000), which would only be a partial solution. Furthermore, the rate of natural production (the majority of the total) would be decreased because of the reduced ambient temperature. Bacteria that oxidize industrially produced ammonia (with carbon dioxide as a carbon source) could be a partial solution as the production of ammonia is only 0.14 Gt/yr (McNutt and Salazar, 2013). Another route is bacteria combining carbon dioxide and hydrogen. The hydrogen could be obtained via electrolysis, reforming methane, or gasifying liquid or solid fuels. However, efficient methods use catalysts, so the analysis of ramping is left to future work.

The most promising chemosynthetic bacteria route is methanotrophic (see Figure 7.5), i.e., utilizing methane as an energy and carbon source. The dry carbohydrate equivalent of global natural gas consumption is 7 Gt/yr (not including natural gas liquids) (IEA, 2012). The carbon

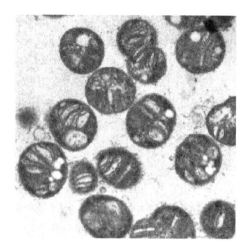

Fig. 7.5. Methylococcus capsulatus *can only digest methane and cannot move.* Image courtesy of Anne Fjellbirkeland (CC).

efficiency of methanotrophs can be up to 62% (Roslev et al., 1997), and we assume 31%. With methane being ~13 kcal/g and ~75% carbon by weight and bacteria being mostly protein (Roslev et al., 1997) (protein is ~4 kcal/g and ~50% carbon), this converts to ~14% energetic efficiency. This yields ~1 Gt of food/yr. These bacteria can operate even down to concentration in equilibrium with atmospheric methane of 1.7 parts per million (ppm) (Roslev et al., 1997). Though they would be exposed to considerably greater methane concentrations, losses could still be made small in covered ponds despite the fact that some methane will exit with the oxygen-depleted air. Possible methods of dewatering the bacteria slurry include centrifugation, flocculation, filtration, gravity sedimentation, flotation, and electrophoresis (Uduman et al., 2010). Raising methane-digesting bacteria on natural gas is already being done commercially (Unibio, 2014). It is being used as an animal feed supplements. But it can also be used as human food with further processing.

Industry could ramp up natural gas production, especially utilizing automobile manufacturing capacity to fabricate natural gas extraction equipment. Bioreactors could grow the bacteria in suspension, so industry would not have to ramp up a growth medium. Furthermore, humans could consume this suspension directly, assuming the bacteria concentration is high enough and industry pasteurizes it and removes important contaminants. In addition, microbes can also consume petroleum (Erdogan and Karaca, 2011) and it is possible to grow mushrooms on lignite (Cohen and Gabriele, 1982) (the type of coal most like peat). Therefore, we assume ramping to full food in 6 months.

Humans could also produce methane by anaerobically digesting biomass, and then feeding this methane to methanotrophs. However, it would be simpler technically and more efficient to have bacteria grow on the biomass directly aerobically. Therefore, we do not consider the methane intermediate as a separate solution.

Therefore, in summary, fossil fuels could provide sufficient feedstocks for a fully chemical (synthetic) or hybrid industrial/biological technique, which would be providing chemicals industrially to chemo-synthesizing bacteria. However, synthetic food production likely cannot be ramped up fast enough.

A further possible solution to using fossil fuels is growing food with artificial light. Sadly, the conversion from fossil fuels to light to plant food is extremely inefficient. The typical conversion of fossil fuel to delivered electricity is approximately 33% efficient (Sims et al., 2003). The conversion efficiency of electricity to visible light is approximately 30% for fluorescent fixtures [a fluorescent fixture is around 65 lumens/W (Myer et al., 2009), and for white light, 100% efficiency is around 220 lumens/W]. Crop plants in both temperate and tropical zones typically do not exceed 1% conversion from solar to food energy (Zhu et al., 2010; Walker, 2009). The theoretical maximum is 9% for micro-algae grown in solar bioreactors, but 3% is more realistic (Wijffels and Barbosa, 2010). Solar energy is only 50% visible (roughly what plants utilize), so the efficiency of plants with artificial light would be correspondingly higher. We assume that the fact that plants utilize lower intensity light more efficiently counteracts the fact that the algae would not be 100% digestible. Therefore, the overall efficiency would be ~0.6% and if all of current electrical consumption could be devoted to photosynthetic food production (there would still be electricity for other uses because of higher duty cycle), this would produce ~5% of the global food requirement. However, since lighting only consumes an order of 10% of electricity, photosynthetic food production would be an order of magnitude smaller. This could be somewhat improved by moving towards LEDs tuned to the photosynthetic absorption optimum. Though industry may be able to double the production rate of lamps (light bulbs) each year, this would only correspond to an order of 10%/yr increase in the actual capacity of lighting. Ramping up LEDs like those now used for commercial Aerogardens (see Figure 7.6) would provide an even

Fig. 7.6. Aerogarden growing herbs. Image courtesy of Sharon Troy (CC).

greater challenge. Therefore, this is not a promising route to full food. It is likely that artificial light would produce very high value products, such as drugs that are not currently synthesized, spices, and food for species with a specialized diet.

All three of these scenarios (synthetic food, chemosynthetic bacteria, and artificial light) have the difficulty of ramping up industrial capacity, except for bacteria oxidizing methane. This latter solution would only require retrofitting to low temperature, pressure, and corrosion bioreactors, and we estimate this could achieve full food in 6 months.

Enzymes and acid can convert lignocellulose (fiber) to sugar (industrial digestion) (Langan et al., 2011). Full food could be produced by industrial digestion in about a year. We have already demonstrated that chipping and grinding capacity would be sufficient for industrial digestion. Therefore, this section covers pretreatment and hydrolysis (breaking down into sugars).

First, there are some combination methods (those that do not require a separate pretreatment step). The concentrated acid technique requires a considerable amount of acid, and because reclaiming the acid would be difficult to ramp up [it is currently just neutralized (Zhu and Pan, 2010)], this technique is less promising. One implementation of the dilute acid technique used 1.4% sulfuric acid with 10% solids (Zheng et al., 2009). Dilute acid can hydrolyze about 60% of "available" cellulose, making up about 12% of the dry mass (Brenner and Rugg, 1986). This would produce approximately 50% as much mass of sugar as sulfuric acid. Very dilute acid (0.07%) or even pure water can hydrolyze nearly all of the hemicellulose (Wyman et al., 2005). Humans do not digest well the most common sugars produced by hemicellulose hydrolysis, except for mannose that is common in softwoods. With a 10% solids loading, hydrolysis of hemicellulose with pure water at 180°C is approximately 65% effective for corn stover (stalks) (Yang et al., 2004). If similar effectiveness could be achieved on softwood, the mannose yield would be ~15% of the dry mass, and this would correspond to approximately 1.7% sugar concentration. The latter technique has the advantages of no acid use, less stringent vessel requirements (i.e., lower pressure, temperature, and corrosiveness), and lower toxicity by-products. However, both of these methods yield sugar too dilute for humans to digest as their sole food source. The typical water consumption per day including free water in food is 2.7 kg (Harte, 1988). Assuming that humans could tolerate twice this, the minimum sugar concentration for full food is ~10%. At 50% boiling efficiency, this would require ~20 Gt/yr of fuel. This is more energy than humans currently consume, but would be feasible with killed trees, which would not be required for other food methods if this were the sole source of food. Multiple effect distillation, reverse osmosis, forward osmosis, etc., are far more efficient, but would require dramatic industrial ramping. Future work is required to assess the retrofitting of existing pressure vessels for the hot water technique. Supercritical water can hydrolyze both hemicellulose and cellulose (Saka, 2006). However, this requires high temperatures and very high pressures; therefore, it again would require dramatic industrial ramping. Therefore, we do not propose acid/water hydrolyses as viable solutions at this time.

Enzymatic hydrolysis achieves much higher yields, but requires a pretreatment step. Pretreatments include dilute acid, steam explosion, high temperature water, ionic liquids, organic solvents (organosolv), fungi, ammonia fiber explosion, strong alkali, concentrated acid (Langan et al., 2011), ammonia recycle percolation, and sulfite pretreatment to overcome recalcitrance of lignocelluloses (Zhu and Pan, 2010). An example of a strong alkali is slaked lime [$Ca(OH)_2$]. Global production of cement was 1.5 Gt/yr in 1995, so extrapolated to 2013, this is 2.7 Gt/yr (Worrell et al., 2001). Cement is ~85% clinker and clinker is ~65% quick lime (CaO) (Worrell et al., 2001). Therefore, production of quick lime just for cement is approximately 1.5 Gt/yr and simply adding water can turn this into 2 Gt/yr of slaked lime. This at 100°C can remove the lignin on 20 Gt/yr of biomass (Wyman et al., 2005), which would be sufficient for all human food even at 8% conversion efficiency. Carbon dioxide can neutralize the leftover slaked lime, and there is a large clean source in the form of natural gas power plant emissions. Furthermore, fungi can ramp up quickly and could even produce food in the form of mushrooms. In addition, industry could use other techniques from the list. Therefore, pretreatment would not be a limiting factor.

Industry currently produces cellulase and hemicellulase (the enzymes that break down cellulose and hemicellulose) by raising fungi and bacteria (Wyman et al., 2005). Normally, industry purifies the enzymes, but it would be possible to kill the organisms, release the enzymes, and put all of this material on the lignocellulose. A number of techniques could achieve this cell disruption, including detergents, lysozymes (enzymes that damage bacterial walls), and ultrasound, but probably the most scalable is grinding. The enzyme-production vessels would not have to handle high temperature, pressure, or corrosion, so retrofits would be straightforward. Also, the fungi and bacteria could be scaled very quickly. Enzyme hydrolysis occurs at less than 100°C (Wyman et al., 2005), so again, vessel retrofits would be simple. In one configuration, resultant glucose (from cellulose) concentrations were 5.5% (Zheng et al., 2009), so with the addition of hydrolysis of hemicellulose, the sugar concentration could be even greater. Increasing the concentration from 5.5% to 10% would require ~3 Gt of fuel for boiling. However, if 50% of calories came from dry food of another source, then sugar could make up ~50% of the calories without any boiling. On the other hand, as is noted for other food types, the removal of water facilitates preservation and reduces shipping energy.

Overall conversion efficiency to sugars with enzymes typically exceeds 50% (Wyman et al., 2005). However, we assume only 30% here, representing nonideal conversion with softwoods (so that the sugar produced by hemicellulose breakdown is digestible by humans) or closer to ideal conversion with other sources of biomass (assuming indigestible sugars from hemicellulose). Therefore, full food would require ~5 Gt/yr of biomass. The only step in the process that causes a considerable time delay is fungal pretreatment, which requires about 2 months with wood chips and lignin-specific fungi (Wan and Li, 2012). The United States mandated enzymatic cellulosic biofuel production to be 1 billion liters of ethanol/yr by 2013 (Venezia and Logan, 2007), so we estimate global production to be 3 billion liters. With ~7 kcal/g, 0.87 kg/L, and 65% efficient from sugar to ethanol (Badger, 2002), this is sugar equivalent to 0.5% of food requirement. Since enzymatic food production is a two-step process (unlike methane digesting bacteria), we assume the retrofit would require 1 year. Analyzing toxicity from processing is future work.

7.4 TREES FOR DINNER? STORED BIOMASS CONVERSION

Generally for the solutions outlined here, we halved the ideal efficiency to take into account practical difficulties. We also generally take the square root of the ideal population growth rate because when there are a large number of offspring, a smaller fraction would be feasible to save.

7.4.1 Beef Steak, Lamb Chops, and Bison Burgers

Ruminants, such as cattle, sheep, and goats, have the ability to digest cellulose (Van Soest, 1994). These already make up a considerable fraction of our calories, but they are limited in their doubling rate. Wood that has been partially decomposed by mushrooms is already being fed to cattle, sheep, and bison (Spinosa, 2008). Other numerous large grazers include horses and deer. Even pigs can digest half of pure cellulose feed (Byerly, 1967). Rabbits can also digest cellulose, and double much faster, but currently have a much smaller initial food production. These all could be partial solutions, but it would likely not save all the people. The ideal growth rate of ruminants with one offspring/yr is ~40%, so with less than ideal food, we estimate a 30% growth rate. Note that this assumption is higher than the customary half used throughout the rest of this analysis because ruminants can digest cellulose and would have high-nitrogen feces supplements. Also, people can expend much effort on each animal because it is large. We leave rabbits to future work. The global consumption of ruminant flesh and milk is 5% of calories (Delgado, 2003). We assume that the loss due to poorer feed is counteracted by techniques to maximize caloric production (e.g., feeding all edible parts of the animals to humans and using horses, deer, etc.), so this would be ~5% of initial global calories from large grazers.

7.4.2 More Mushrooms

A log 1 m long and 0.1 m in diameter will produce approximately 1 kg of wet mushrooms over 4 years (Hazeltine and Bull, 2003). Because the dry density of wood is 0.5 g/cm^3 (Ragland et al., 1991), and mushrooms are ~10% dry weight (Chang and Miles, 2000), this yields approximately 2% caloric efficiency. It also may be feasible to separate the mushroom mycelia ("roots") for human ingestion, increasing the food supply. Note that 10% dry weight is feasible from a water ingestion perspective even if the mushrooms are not dried. Thus, outdoor mushroom growth on logs is ~2% efficient or closer to 1% because of nonideal logs, pests, and inexperienced human error. As 10 cm diameter logs are consumed in ~4 years, only about 70% of the mass of trees cut down could be consumed in 4 years. Thus, ~700 Gt of dry wood would have to be prepared in 1 year, which does not appear to be feasible from a wood supply perspective. Ramping is not a constraint because of billions of spores per mushroom. Therefore, mushrooms can supply only roughly 25% of human food for the longer term (we covered the short-term use of mushrooms in Chapter 5). The 20°C crisis would have lower longer-term mushroom production because of less biomass in nonfreezing areas.

7.4.3 Beetles

The caloric efficiency of beetles digesting xylem that is primarily cellulose is 12% (Benner et al., 1988). In the general case of digesting lignocellulose with the real-world issue of pests, a more realistic number is probably ~6%. For most cellulose-digesting beetles, the wood would need to be softened, and ideally have the lignin removed. Grinding capacity is insufficient, though chipping and industrial lignin removal would be feasible with some ramping as discussed in Chapter 6. Another option is partial digestion by fungi. Since white rot fungi (including mushrooms) preferentially digest lignin (Belewu and Belewu, 2005), if they consume 20% of the calories of the wood at 5% efficiency, this would leave ~80% of the calories in the wood (similar to the industrial lignin removal scenario). If the cellulose-digesting beetles have an efficiency of 6%, this would be overall ~5% conversion efficiency from wood to animal.

Though many bark beetles are not cellulose digesting, we assume that bark beetles provide an estimate of all cellulose-digesting beetles. Ten million hectares in the Western United States and British Columbia are affected by bark beetles (Bateman, 2012), so assuming 200% more area globally, this is ~1% of forest area. This corresponds to about 3 Gt of above-ground mass. Since the cambium (inner bark) is approximately 1 mm thick (Rothwell, 2008), this is about 2% of the mass of the typical 20 cm log diameter (including branches). If beetles have consumed this on average over the affected area in 5 years at 6% efficiency, this is ~0.001 Gt dry beetle biomass/yr. We assume that the beetle biomass production rate in the tropics cooled by the disaster would be similar to the current production rate in the mid-latitudes. A closely related beetle lays ~16 eggs and has ~two generations/yr (Weber and McPherson, 1983). Therefore, the ideal growth rate would be a factor of 64/yr, so to remain conservative, we assume the square root of this.

In summary, the initial stock of cellulose-digesting beetles is sufficient to provide ~0.05% of human food and with a realistic growth rate for the 10°C crisis, this is a factor of 8 growth in 1 year. Therefore, it would take ~4 years to reach full human food. With a 20°C crisis, it is unlikely that the beetles could reproduce; however, chipping the logs and moving them indoors could provide a workaround.

7.4.4 Rats or Chicken?

In natural ecosystems, bacteria make lignocellulose available to non-cellulose digesters (Benner et al., 1988). An example of this is fish eating partially decomposed leaves to get the digestible calories in the bacteria. Similarly, we propose to use fungi and bacteria to process fiber for digestion by rats and chickens. Humans already consume rats, and there is evidence that rats can digest cellulose (Johnson et al., 1960). Rats have some ability to digest cellulose [it varies depending on the type of plant and ranges from 0.5% to 20.84% (Keys et al., 1969)]. It should be noted that the presence of fiber that cannot be digested decreases the energy yield from eating food and thus some caution is needed for using high-fiber-material-conversion efficiencies.

Having rats eat wood that has had the lignin reduced by fungi and most of the cellulose and hemicellulose converted into bacteria (with fertilization) would have an overall efficiency of ~4%. We can give the waste from growing mushrooms on logs or on thin biomass (from the

Fast Food Chapter 5) to bacteria. However, it would be advantageous to have the bacteria distributed evenly to allow uniform consumption (to maximize the available calories in the bacteria) and allow the administering of fertilizer. This would require chipping of the logs, and there would be sufficient capacity as we showed in Chapter 6, especially because the fungi would soften the wood. If fungi/bacteria digested all components of the wood at equal rates to 95% consumption, and the bacteria were 95% digestible by noncellulose digesters, and the bacteria have the ideal carbon (and approximately caloric) efficiency of 45% (Benner et al., 1988), fiber would make up only 15% of the final dry weight. However, the fiber percentage would likely be higher, so this requires further research. If the decomposition was at half ideal efficiency (because of pests, nonideal fertilizer, and nonideal substrates) and it leaves 15% of the original wood calories undigested, this would be an apparent 34% efficiency. The ideal rat caloric conversion efficiency is 35% at weanling age (Byerly, 1967). Since typical bacteria have a very low carbohydrate percentage, the rats could have a response similar to humans with the Atkins diet of consuming body fat instead of weight maintenance or gain. However, with the limited ability to digest fiber, this should provide adequate carbohydrates to the rat. Furthermore, if the rats are eating the bacteria continuously, the bacteria would have turned some of the cellulose into sugar outside the bacteria's bodies. Bacteria with considerable carbohydrate storage would be advantageous because they would reduce the nitrogen requirements of the bacteria and possibly increase conversion efficiency, but this requires further work. If rats achieved one-third of the ideal conversion efficiency due to all the fiber and other issues, the overall efficiency would be ~4%. We use the rat calories as a proxy for the human-digestible food, because bacteria or animals could digest the indigestible portions of the rats, and because humans could get some calories by having other organisms digest the rat waste. Finally, it is possible that the rats or chickens could pick out the mycelia and just eat that, and then enzymes, ruminants, or bacteria could consume the rest. High efficiency would require a considerable amount of fertilizer with conventional low-carbohydrate bacteria (unless there were considerable nitrogen fixation), but the supply is likely to be adequate as shown below.

Nitrogen is the focus for fertilizer requirements because it is generally the shortest in supply. The current global production of fixed nitrogen (mostly for fertilizer) is 0.14 Gt/yr (McNutt and Salazar, 2013). The supply of nitrogen from domesticated animal excrement is 0.1 Gt/yr (Van der Hoek, 1998). This could increase more than an order of magnitude as animal production is ramped up and as animal diets become higher in nitrogen (e.g., eating bacteria). Much of this excrement could be fed to other species, but still most of the nitrogen eventually would be available as fertilizer. Drying and transporting all this natural fertilizer considerable distances would likely not be feasible, but farmers could utilize it locally. The consumption of bacteria (whether grown on thin biomass, trees, or methane) is high overall efficiency from feedstocks to human calories, so the fertilizer requirement is low. Enzyme-produced food may require some fertilizer to produce the enzyme, but the sugar production would not require fertilizer, so the overall fertilizer requirement would be small due to the high efficiency. There is adequate nitrogen in the biomass to support mushroom and cellulose-digesting beetles without any addition of fertilizer. Ruminants and rabbits do not require a bacteria intermediary, so the nitrogen requirement would be moderate. Therefore, the critical cases are rats and chickens eating mostly digested biomass. As stated before, the ideal conversion efficiency of bacteria based on full utilization of nitrogen in wood is approximately 10%. The ideal conversion efficiency of bacteria unconstrained by nitrogen is 46%, and we assume half that. Then, we assume a fertilizer utilization of 90% because the scenario is chipped wood indoors, and run-off and volatilization (conversion to gas) can be tightly controlled. This would correspond to approximately 60% of the nitrogen in the bacteria, which must be added as fertilizer. This yields 200% more than current artificial nitrogen production. However, with high-protein diets, animals excrete the majority of the nitrogen in the feed (Byerly, 1967). Therefore, since there would be a relatively small delay in recycling of the nitrogen, fertilizer supply would be adequate. [While we are discussing fertilization we should discuss the oceans. There have been experiments with iron fertilization of the ocean. They have been disappointing with respect to sequestering carbon, but this means that the nutrients are recycled, with only ~10% of the expected carbon in the excrement falling to the ocean floor (Watson et al., 1994). This indicates that with macronutrient fertilization, the nutrients would likely be mostly recycled, meaning the fertilizer requirement would be moderate.] Furthermore, doubling industrial production of fertilizer each year as we discuss below would ameliorate the issue.

It is possible that the rats could provide 5 years of food if they consume the entire pretreated log. Once mushrooms partially digest the logs and the wood chippers chip the logs, people would move the logs indoors. This would reduce the problem of competing organisms. People could apply fertilizer with near unity conversion into the bacteria because people would recycle the drainage and control the fertilizer concentrations to limit volatilization.

We would have a good start with rats. They outnumber humans (Langton, 2007), so we assume that humans would be able to catch a number of rats equal to the human population. An adult rat weighs approximately 90 g dry (Schulte-Hostedde et al., 2001) and we assume a calorie density of carbohydrates (similar to fish). They achieve half of adult rat weight at the average of sexual maturity of six weeks, and near adult weight after 12 weeks (Sengupta, 2011). Finally, we adjusted the total calories downward by 20% to represent nonideal food. These statistics are with lab rats. For population growth, we could either adjust the lab rat numbers downward or use wild numbers. We take the latter approach here. With a sexual maturity time of 18 weeks (Davis, 1949) and giving birth to 7–9 offspring every 70 days (Storer and Davis, 1953), the annual growth rate of rats would be a factor of ~50. Photoperiod should not be an issue.

In summary, the fertilizer supply should be adequate and rats could provide all food since they could consume the entire pretreated log because of chipping. The current rat production is order 0.1% of our food requirements. With the square root of the ideal growth rate, this would take ~2 years to ramp up to 100% of human food. Furthermore, other rodents would provide additional food.

Although perhaps more appetizing, the inherent efficiency of chickens is very similar to that of rats (Byerly, 1967). However, chickens have very little ability to digest cellulose, so the resultant mostly decomposed wood would have to be low in fiber. Chickens can eat their own excrement, so it might be possible for them to eat raw decomposed wood. If not, pasteurization would make the wood safe. In addition, chickens may require another source of carbohydrate. Additional sources of carbohydrates for chicken feed include industrially produced sugar, sugar outside the bacteria's bodies, mushroom mycelia, and leaf extraction. Global consumption of poultry and eggs is ~2% of calories (Delgado, 2003). We assume that the loss due to poorer food counteracts wild birds, reduction in wastes, and techniques to maximize caloric production (e.g., feeding all edible parts of the birds to humans).

With a sexual maturity time of 5 months and laying an egg every 2.5 days (Amira, 2008), the annual growth rate of chickens is a factor of ~ 500. This is already slower than some intensive operations, and considerable effort can be expended because of the medium size of these animals, so we assume a population growth rate that is the three-fourths power of the ideal, or a factor of 100/yr.

Typical daylight is 5000 lumens per square meter (illumination), and livestock require about 150 lumens per square meter to perceive daylight (Collier and Collier, 2011). Therefore, the disaster would have to block more than 97% of the light to affect reproduction via photoperiod, which is quite unlikely. However, the nearly constant 12-h photoperiod in the tropics may not be optimal for reproduction or growth. We absorb this effect in the conservatism from the ideal reproduction rate and efficiency.

Depending on the severity, a crisis may require some relocation of people. However, we consider the worst-case scenario from a distribution perspective of no relocation. The refrigeration capacity would be relatively small compared to the required flow of food, but drying is feasible and would obviate refrigeration. In addition, transporting the dried food from the tropics would take a minority of current capacity. The critical case for drying is industrial digestion, but it would be feasible.

7.5 WHAT WILL PROBABLY NOT WORK: SHIPWORMS, TERMITES, GRIBBLES, EARTHWORMS, AND REPTILES

Shipworms (a type of mollusk) have the ability to digest wood, have on the order of 1,000,000 eggs and mature in ~ 3 months (Ho, 2012), making them a viable candidate. However, raising marine organisms is challenging because of the difficulty of removing waste products. This precludes indoor production, but it may be possible to grow the shipworms over the tropical continental shelf. They would likely only work for the 10°C crisis, because low temperatures inhibit reproduction. Future work includes locating a conversion efficiency of shipworms as it appears that no one has bothered to determine this before.

Termites are also unlikely candidates because they are also sensitive to low temperatures. In addition, termites are difficult to rise in captivity

(Leuthold et al., 2004). Therefore, a considerable fraction of human food is unlikely to come from termites.

Gribbles (a type of crustacean) also have the ability to digest wood, but have a much slower theoretical population growth rate than ship-worms [order 10 eggs each time (Thiel, 2003)], and much smaller initial stock than cellulose-digesting beetles, so ramping would not be feasible.

Some earthworms can digest cellulose (Nozaki et al., 2009). Earth-worms have a large initial stock and rapid potential population growth rate. Humans would probably have to transport earthworms from the mid-latitudes to the tropics because the temperature of the tropics would be lower. It is not clear that these earthworms could accommodate the change in soil. Furthermore, since earthworms have limited ability to break up larger particles, competing organisms would probably consume most of the available calories in wood before earthworms could ingest it. Therefore, we do not consider earthworms here as a major source of human calories.

There are also some reptiles and amphibians that digest cell walls (Mackie and White, 1997). However, with a relatively small stock and not exceptional growth rate, they do not appear to be promising food sources.

There are other noncellulose digesting organisms that could be ramped up quickly, such as ants. However, since the farming of chickens is well developed and high efficiency and there is a taste and cultural acceptability advantage, chicken would probably be preferable. Humans could also raise pigs, but they cannot ramp as fast as chickens, so pigs would only be a partial solution.

Thus, we do not believe that shipworms, termites, gribbles, earth-worms, reptiles, amphibians, or noncellulose digesting insects are prom-ising solutions.

7.6 A BANQUET

In summary, we have developed several full food solutions for the global catastrophes outlined in this book. The ramp rates of all the solutions are provided in Figures 7.7 and 7.8 for the 10 and 20°C crises, respectively. Figure 7.8 shows the approximate food supply over time from the most promising solutions presented in this Chapter. Note how some of the

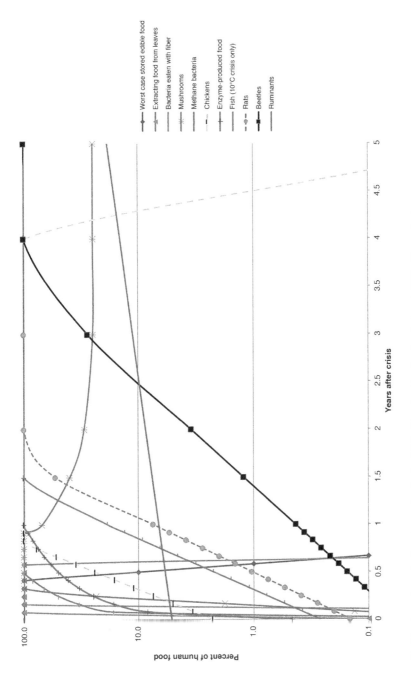

Fig. 7.7. Approximate food supply over time for the 10 °C crisis ordered by when they first reach 100% of human food.

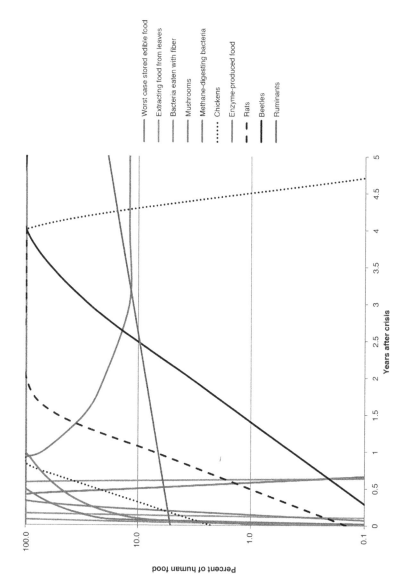

Fig. 7.8. Approximate food supply over time for the 20 °C crisis ordered by when they first reach 100% of human food.

supplies drop off quickly with time (e.g., stored food). The sources are ordered by when they first reach 100% of humanity's food needs. Many of these food sources are not independent of each other and one cannot simply add them, although it is clear from the figures that much more than 100% human food is available for 5 years with these solutions. The independent groups are stored food, methane bacteria, fast food (generally nonwoody biomass sourced), slow food with animals (wood sourced), wood converted to bacteria and enzyme-produced food (the latter two are high efficiency foods, so little competition for biomass). Even within these groups, assuming biomass supply is adequate, there are some independent routes to full food supply. There is very large uncertainty in these numbers, so the primary purpose is to illustrate the magnitudes and trends. The least certain are rats and chickens, dotted.

These solutions can work together in an industrial ecology food web, as seen in Figure 7.9, which shows the most important food solutions

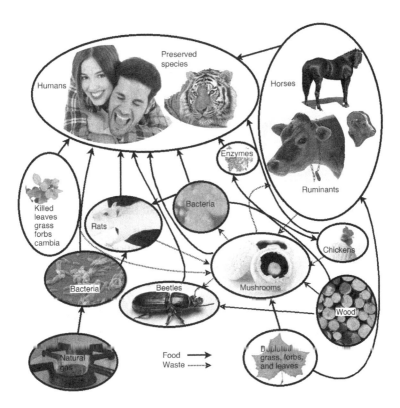

Fig. 7.9. Most important food sources and energy flows of the transient detritus food web. Cambia are inner barks and forbs are nonwoody, nongrass plants (fish based on photosynthesis are omitted).

and energy flows. Figure 7.9 can be thought of as a transient detritus (dead material) food web (plus natural gas).

7.7 MOST EXTREME CATASTROPHES

Even in the most extreme disasters, such as a very large asteroid or comet that could result in the burning of nearly all above-ground vegetation, there may still be some viable routes to feed everyone.

Very large impacts would increase atmospheric temperature considerably immediately after impact (Melosh, 1990). Therefore, this would require some advance warning, but this would generally be the case because of the size of the impactor. Possible solutions to the temperature increase for buildings include consolidating people into fewer buildings, rejecting heat to the ground, using air compressors for cooling (compressing air, letting it cool down, and then decompressing to go to low temperatures), and relocating to cooler areas such as caves, existing mines, or even onto ice. Also, there would be intense thermal radiation caused by rock particles (Melosh, 1990), so insulating on the outside of buildings would protect from this radiation and facilitate longer-term building cooling. Furthermore, industry could retrofit vehicles for the higher temperatures. Toxic gases could be a problem globally, but possible solutions include filtration (e.g., activated carbon), destruction via catalysts, and consumption by bacteria.

As for food, stored food could be removed from the high-destruction areas and protected. Methane-digesting bacteria would still be promising. Also, enzymatic conversion is efficient, so the feedstock requirement is small. Vegetation that naturally did not burn, biomass that humans protected, or peat (its high water content would protect it and it still generally contains considerable amounts of cellulose and hemicellulose (Andriesse, 1988)) could meet this requirement.

It is conceivable that it takes more than one decade to solve the crop-killing scenarios. One route to extending the existing vegetation supply is ramping up industrial digestion that is considerably more efficient than biological methods. Again, methane-digesting bacteria would be feasible. Furthermore, with the greater lead time, it is possible that industry could ramp up synthetic food production to provide all of human food. The energy source could be either fossil fuel or renewable energy, because these scenarios would not involve blocking the Sun.

REFERENCES

Amira, E.E.-D., 2008. The relationship between age at sexual maturity and some productive traits in local chickens strain. Egypt Poultry Sci. 28, 1253–1263.

Andriesse, J.P., 1988. Nature and management of tropical peat soils. FAO Soils Bull. 59, Rome.

Badger, P.C., 2002. Ethanol from cellulose: A general review. In Trends in New Crops New Uses. ASHS Press, Alexandria, VA, pp. 17–21.

Bateman, M., 2012. Insect outbreaks kill forests and release carbon. Live Science. http://www.livescience.com/18797-beetle-outbreaks-forests-carbon-nsf-bts.html

Bates, T.S., Lamb, B.K., Guenther, A., Dignon, J., Stoiber, R.E., 1992. Sulfur emissions to the atmosphere from natural sources. J. Atmos. Chem. 14, 315–337.

Belewu, M.A., Belewu, K.Y., 2005. Cultivation of mushroom (*Volvariella volvacea*) on banana leaves. Afr. J. Biotechnol. 4, 1401–1403.

Benner, R., Lay, J., K'nees, E., Hodson, R.E., 1988. Carbon conversion efficiency for bacterial growth on lignocellulose: Implications for detritus-based food webs. Limnol. Oceanogr. 30, 1514–1526.

Bostrom, N., Cirkovic, M.M. (Eds.), 2008. Global Catastrophic Risks. Oxford University Press, New York.

Brenner, W., Rugg, B., 1986. High temperature dilute acid hydrolysis of waste cellulose. Continuous and batch processes. United States Environmental Protection Agency. EPA/600/S2-85/137.

Byerly, T.C., 1967. Efficiency of feed conversion. Science 157, 890–895.

Chang, S.-T., Miles, P.G., 2000. Edible mushrooms and their cultivation. CRC Press.

Collier, R.J., Collier, J.L., 2011. Environmental Physiology of Livestock. John Wiley & Sons, New Delhi, India.

Cohen, M.S., Gabriele, P.D., 1982. Degradation of coal by the fungi *Polyporus versicolor* and *Poria monticola*. Appl. Environ. Microbiol. 44, 23–27.

Corbett, J.J., Koehler, H.W., 2003. Updated emissions from ocean shipping. J. Geophys. Res. 108, 4650.

Davis, D.E., 1949. The weight of wild brown rats at sexual maturity. J. Mammalogy 30, 125–130.

Delgado, C.L., 2003. Rising consumption of meat and milk in developing countries has created a new food revolution. J. Nutr. 133, 3907S–3910S.

Erdogan, E., Karaca, A., 2011. Bioremediation of crude oil polluted soils. Asian J. Biotechnol. 3, 206–213.

Eurofish, 2012. Overview of the world's anchovy sector and trade possibilities for Georgian anchovy products.

Fraser, E., 2014. Can we feed everyone? CNN. http://us.cnn.com/2014/10/15/opinion/fraser-world-food-day/index.html

Harte, J., 1988. Consider a Spherical Cow: A Course in Environmental Problem Solving. Sausalito, CA, University Science Books.

Hartman, K.J., Brandt, S.B., 1995. Estimating energy density of fish. Trans. Am. Fisheries Soc. 124, 347–355.

Hazeltine, B., Bull, C., 2003. Field Guide to Appropriate Technology. San Francisco, Academic Press.

Ho, M. 2012. Teredo navalis. Available: http://animaldiversity.ummz.umich.edu/site/accounts/information/Teredo_navalis.html.

Hudlicky, T., Entwistle, D.A., Pitzer, K.K., Thorpe, A.J., 1996. Modern methods of monosaccharide synthesis from non-carbohydrate sources. Chem. Rev. 96, 1195–1220.

IEA, 2012. World Energy Outlook 2012. International Energy Agency.

Johnson, R.B., Peterson, D.A., Tolbert, B.M., 1960. Cellulose metabolism in the rat. J. Nutr. 72, 353.

Keys, J.E.J., Van Soest, P.J., Young, E.P., 1969. Comparative study of the digestibility of forage cellulose and hemicellulose in ruminants and nonruminants. J. Animal Sci. 29, 11–15.

Kuwahara, V.S., Toda, T., Hamasaki, K., Kikuchi, T., Taguchi, S., 2000. Variability in the relative penetration of ultraviolet radiation to photosynthetically available radiation in temperate coastal waters, Japan. J. Oceanogr. 56, 399–408.

Langan, P., Gnanakaran, S., Rector, K.D., Pawley, N., Fox, D.T., Chof, D.W., Hammelg, K.E., 2011. Exploring new strategies for cellulosic biofuels production. Energy Environ. Sci. 4, 3820–3833.

Langton, J., 2007. Rat: How the World's Most Notorious Rodent Clawed its Way to the Top. Key Porter Books Limited, Canada.

Leuthold, R.H., Triet, H., Schildger, B., 2004. Husbandry and breeding of Giant African Termites (*Macrotermes jeanneli*) at Berne Animal Park. Der Zoologische Garten 72, 26–37.

Mackie, R., White, B., 1997. Gastrointestinal Microbiology, Vol. 1: Gastrointestinal Ecosystems and Fermentations. Springer, New York.

McNutt, M.K., Salazar, K., 2013. Mineral commodity summaries 2013, 116.

Melosh, H.J., Schneider, N.M., Zahnle, K.J., Latham, D., 1990. Ignition of global wildfires at the Cretaceous/Tertiary boundary. Nature 343, 251–254.

Mercado, L.M., Bellouin, N., Sitch, S., Boucher, O., Huntingford, C., Wild, M., Cox, P.M., 2009. Impact of changes in diffuse radiation on the global land carbon sink. Nature 458, 1014–1018.

Myer, M.A., Paget, M.L., Lingard, R.D., 2009. Performance of T12 and T8 fluorescent lamps and troffers and LED linear replacement lamps.

Nellemann, C., Hain, S., Alder, J., 2008. In Dead Water: Merging of Climate Change with Pollution, Over-Harvest, and Infestations in the World's Fishing Grounds. United Nations Publications, Norway.

Nozaki, M., Miura, C., Tozawa, Y., Miura, T., 2009. The contribution of endogenous cellulase to the cellulose digestion in the gut of earthworm (*Pheretima hilgendorfi*: Megascolecidae). Soil Biol. Biochem. 41, 762–769.

Pacala, S., Socolow, R., 2004. Stabilization wedges: Solving the climate problem for the next 50 years with current technologies. Science 305, 968–972.

Pauly, D., 1996. One hundred million tonnes of fish, and fisheries research. Fish. Res. 25, 25–38.

Pearce, J.M., 2008. Thermodynamic limitations to nuclear energy deployment as a greenhouse gas mitigation technology. Int. J. Nucl. Governance, Econ. Ecol. 2, 113–130.

Pinet, P.R., 1992. Oceanography: An Introduction to the Planet Oceanus. West Publishing Company.

Ragland, K.W., Aerts, D.J., Baker, A.J., 1991. Properties of wood for combustion analysis. Bioresource Technol. 37, 161–168.

Roslev, P., Iversen, N., Henriksen, K., 1997. Oxidation and assimilation of atmospheric methane by soil methane oxidizers. Appl. Environ. Microbiol. 63, 874–880.

Rothwell, M., 2008. Selection of tree species for cambium consumption by the Bornean orangutan (Pongo pygmaeus wurmbii), Dissertation Selwyn College. 2008.

Runge, C.F., Senauer, B., 2007. How biofuels could starve the poor. Foreign Affairs, 41–53.

Saka, S., 2006. Recent progress in supercritical fluid science for biofuel production from woody biomass. For. Stud. China 8, 9–15.

Schulte-Hostedde, A.I., Millar, J.S., Hickling, G.J., 2001. Evaluating body condition in small mammals. Can. J. Zool. 79, 1021–1029.

Sengupta, P., 2011. A scientific review of age determination for a laboratory rat: How old is it in comparison with human age? Biomed. Int. 2, 81–89.

Sims, R.E., Rogner, H.H., Gregory, K., 2003. Carbon emission and mitigation cost comparisons between fossil fuel, nuclear and renewable energy resources for electricity generation. Energy Policy 31 (13), 1315–1326.

Smith, P.E., 1985. Year-class strength and survival of O-group clupeoids. Can. J. Fish. Aquat. Sci. 42, s69–s82.

Spinosa, R., 2008. Fungi and sustainability. Fungi 1, 138–143.

Storer, T.I., Davis, D.E., 1953. Studies on rat reproduction in San Francisco. J. Mammalogy 34, 365–373.

Thiel, M., 2003. Reproductive biology of *Limnoria chilensis*: Another boring peracarid species with extended parental care. J. Nat. Hist. 37, 1713–1726.

Uduman, Ni., Qi, Y., Danquah, M.K., Forde, G.M., Hoadley, A., 2010. Dewatering of microalgal cultures: A major bottleneck to algae-based fuels. J. Renew. Sustain. Energy 2, 012701.

Unibio, 2014. What is Uniprotein(R)?http://www.unibio.dk/?page_id=47

United Nations Children's Fund (UNICEF), 2006. The State of the World's Children 2007. Vol. 7. (UNICEF).

Van der Hoek, K.W., 1998. Nitrogen efficiency in global animal production. Environ. Pollut. 102, 127–132.

Van Soest, P.J., 1994. Nutritional Ecology of the Ruminant. Cornell University Press, Ithaca, NY, USA.

Venezia, J., Logan, J., 2007. Weighing U.S. energy options. The WRI bubble chart.

Wallace, J., Hobbs, P., 1977. Atmospheric Science. Academic Press, New York.

Walker, D.A., 2009. Biofuels, facts, fantasy, and feasibility. J. Appl. Phycol. 21, 509–517.

Wan, C., Li, Y., 2012. Fungal pretreatment of lignocellulosic biomass. Biotechnol. Adv. 30, 1447–1457.

Watson, A.J., Law, C.S., Scoy, K.A.V., Millero, F.J., Liddicoat, M.I., Wanninkhof, R.H., Barber, R.T., Coale, K.H., 1994. Minimal effect of iron fertilization on sea-surface carbon dioxide concentrations. Nature 371, 143–145.

Weber, B.C., McPherson, J.E., 1983. Life history of the Ambrosia Beetle *Xylosandrus germanus* (Coleoptera: Scolytidae). Ann. Entomol. Soc. Am. 76, 455–462.

Wijffels, R.H., Barbosa, M.J., 2010. An outlook on microalgal biofuels. Science 329, 796–799.

World Health Organization, 2000. Air Quality Guidelines for Europe, second edition. WHO Regional Publications, Eurepean Series, No. 91. WHO, Copenhagen.

Worrell, E., Price, L., Martin, N., Hendriks, C., Meida, L.O., 2001. Carbon dioxide emissions from the global cement industry. Annu. Rev. Energy Env. 26, 303–329.

Wyman, C.E., Decker, S.R., Himmel, M.E., Brady, J.W., Skopec, C.E., Viikari, L., 2005. Hydrolysis of cellulose and hemicellulose. In Polysaccharides: Structural Diversity and Functional Versatility. Marcel Dekker, Inc, New York, 995–1033.

Yang, B., Gray, M.C., Liu, C., Lloyd, T.A., Stuhler, S.L., Converse, A.O., Wyman, C.E., 2004. Unconventional relationships for hemicellulose hydrolysis and subsequent cellulose digestion. In ACS Symposium Series. Oxford University Press, Cary, NC, 100–125.

Zheng, Y., Pan, Z., Zhang, R., Wang, D., 2009. Enzymatic saccharification of dilute acid pretreated saline crops for fermentable sugar production. Appl. Energy 86, 2459–2465.

Zhu, X.-G., Long, S.P., Ort, D.R., 2010. Improving photosynthetic efficiency for greater yield. Annu. Rev. of Plant Biol. 61, 235–261.

Zhu, J.Y., Pan, X.J., 2010. Woody biomass pretreatment for cellulosic ethanol production: Technology and energy consumption evaluation. Bioresource Technol. 101, 4992–5002.

CHAPTER 8

Practical Matters: Energy, Water, Nutrition, Taste, Biodiversity, & Cooperation

8.1 PRACTICAL MATTERS

This Chapter covers the practical matters concerning the technical viability for the solutions provided in Chapter 7, specifically those related to energy, water, nutrition, taste, biodiversity, and most importantly cooperation.

8.2 ENERGY IN THE SUN-OBSCURING CRISES

The first part of the energy equation is the reduction in supply due to the crises. We discussed the loss of solar energy in Chapter 3. There is obviously a major reduction in solar photovoltaic power and also likely to be a dramatic reduction in wind power, but this is also order 1% of conventional primary energy (this percent is likely to increase and wind power will be further discussed below). Subsequently, we analyze the increases in energy demand due to solving the crises and conservation methods so that supply can meet demand.

With an important reduction in solar energy, average wind speeds would likely fall considerably because wind is solar-driven. Also, the wind pattern would change, likely meaning that the placement of wind turbines would no longer be optimal, further reducing wind power potential. With wind power being proportional to the cube (third power) of the wind velocity, the resulting wind power would be much smaller than current production. Current global wind power production is ~3% of electricity. Since electricity is 38% of global primary energy (Sims et al., 2003; IEA, 2012), this means wind power makes up ~1% of global primary energy (crediting the wind electricity at the typical power plant consumption of primary energy per electricity production). Other uses of wind energy are negligible, although reduced wind speeds would actually reduce building heating and air conditioning energy consumption.

Feeding Everyone No Matter What. http://dx.doi.org/10.1016/B978-0-12-804447-6.00008-5

In the sun-obscuring crises, temperatures would fall, requiring more building heating, while reducing air-conditioning energy use. If building heating capacity were insufficient and insulation could not be added quickly enough, some relocation may be required from the areas suffering the greatest temperature loss (typically centers of continents) to warmer areas (e.g., coasts), caves, or existing mines. Farming energy use would be saved, but this would be more than compensated by the energy to fell trees to provide the food solutions from the last Chapter, at least in the first year. In extreme scenarios, freight energy could be doubled, meaning an increase of global primary energy of approximately 9% assuming U.S. values of freight of 9% of primary energy use apply globally (U.S. Department of Energy, 2011). The energy to dry the food would be dependent on whether the foods were primarily mushrooms, industrial digestion or leaf extraction (~90% water), animals (~70% water), or bacteria (~50% water). If the drying facilities were 50% efficient based on the heat of vaporization of water, the critical case would be 2 Gt of fuel/yr for mushrooms. If this were fossil fuel, it could make a considerable impact on consumption of the current 10 Gt/yr. However, burning local wood would be a simple solution.

Overall, the increase in energy demand should only be around 10%. While disruption of life as we currently know it would be inevitable and possibly profound, here we assume that life would mostly continue normally to estimate energy savings potential. One promising way of conserving current energy use is by reducing noncommuting personal trips, because light-duty vehicle energy use is 17% of total U.S. primary energy (U.S. Department of Energy, 2011), and commuting is 27% of U.S. personal vehicle miles traveled (U.S. Department of Transportation, 2001). Other energy-saving options that individuals can implement rapidly include:

- lowering heating set points;
- raising cooling set points;
- sealing ducts and exteriors;
- turning off lighting and appliances not in use;
- rack drying of laundry;
- taking shorter duration showers;
- unplugging phantom loads and redundant appliances (e.g., second refrigerators);

- adding a clear polymer layer to single-pane windows; and
- reducing commuting energy use through driving slower, telecommuting, carpooling, taking transit, etc.

Individuals could also save indirect energy use by buying fewer non-essential goods (reducing manufacturing and accompanying shipping savings). Though many of these energy conservation measures do not maintain consumer performance, it is interesting to note that many energy efficiency measures (EEMs) do maintain consumer performance and have high returns on investment (ROI) now (see information Box 8.1).

Furthermore, society could implement some amount of structural changes quickly, such as reduced speed limits and discontinuing the use of energy inefficient vehicles, buildings and factories, and insulating un-insulated buildings, hot water heaters, pipes, ducts, etc. Consolidating people in fewer buildings may save energy, even if people grow food in those unoccupied buildings, as some organisms tolerate lower temperatures (e.g., mushrooms). An example of considerable conservation of electricity resulted when the hydropower plant suffered an outage in Juneau, Alaska, U.S. Electricity had to be replaced with expensive petroleum generation, and electricity consumption fell by more than 40% (Harris, 2011).

We focused this analysis on total energy. In reality, there would not be a perfect match of the conservation ability for each fuel and the increased requirement for each fuel for disaster response. However, there is considerable substitutability of fuels, such as in home heating and electricity generation. Furthermore, biomass that is less valuable for food production, such as wood outside the tropics, could substitute for some fossil fuel.

Artificial-light photosynthesis is unlikely to demand much more than 4% of primary energy because of lamp limitation (see Chapter 7.2).

The psychological aspects of losing the sun could be analogous to the seasonal affect disorder that many people acquire in the winter – perhaps more intense because of the radically altered society needed to maintain survival. One treatment for this is spending a considerable amount of time in high light levels. This would be energy intensive, unless people could use stray light from plant-growing operations. This might be a good way to balance artificial light farming duties with the benefits of full spectrum light therapy.

Information Box 8.1

Working as engineers we have found that most energy efficiency measures (EEMs) make economic sense but involve an initial investment, which is made back with a lower recurring cost than the standard devices or processes. Telling people the payback time often discourages rational investment. Payback time is simply the capital cost divided by the yearly savings from implementing an EEM and it is what most people use to make decisions. For example, if you are considering putting solar cells on your home that save you $1,000 a year in offset electric costs but they cost $10,000 up front, many people will not make the investment because the simple payback time is 10 years. That does seem like a long time. However, it is far better to convert the payback time into a return on investment to make a proper financial decision. To do this you need the lifetime of the investment – or in this case the lifetime of the solar cells. Solar cells are solid-state devices that will likely last longer than you will – but generally have warranties for 20–30 years (Branker et al., 2011). If we use 20 years and the graph (Pearce et al., 2009) in Figure 8.1 we can convert the simple payback time of 10 years (find it on the *x*-axis) into the ROI by tracing up to the lifetime (20 years on the y axis) – which turns out to be about 8% return on investment. Knowing you will make 8% on your money for every year for the next 20 years is currently considered a very good investment as current similarly secure returns are remarkably low and are not keeping up with inflation. Perhaps even more attractive, is that this investment would be a tax-free 8% return. The easy-to-understand method of translating payback time and the more familiar return on investment is valid for the majority of cases of EEMs.

Some EEMs are many relatively minor improvements that would cut energy use while driving high ROIs include:

- energy efficient lighting retrofits (e.g., LEDs);
- timers, occupancy sensors, and CO_2 sensors to adjust lighting and HVAC based on occupancy of buildings;
- thermal insulation improvement and advanced windows;
- increase recycling of all materials;
- convert to high-efficiency appliances (e.g., ENERGY STAR);
- improve vehicle fleet fuel efficiency by retiring inefficient vehicles.

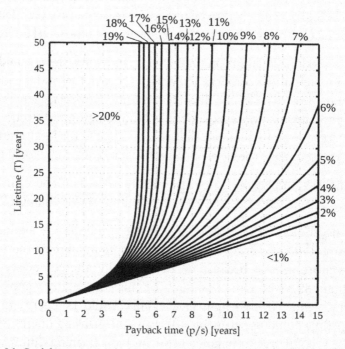

Fig. 8.1. Graph for converting payback time to ROI using lifetime of an energy efficiency measure.

Wind power has grown at 25%/yr (Kammen, 2006), and as photovoltaic technology has dropped in cost to often out-compete conventional supplies (Branker et al., 2011) PV has grown at over 45% (Kenny et al., 2010). Specific types of photovoltaic technologies have increased by more than double that (e.g., thin-film photovoltaic plants have increased by over 100%) (Branker et al., 2011) so the crisis would allow even higher industrial growth rate. Therefore, we assume energy production could increase at 100%/yr, meaning energy supply would be adequate. We assume this industrial ramping would apply to other sectors as well.

8.3 WATER

Global precipitation is 1.12 m/yr (Legates and Willmott, 1990). For the nuclear winter scenario of 9°C temperature reduction, the precipitation reduction was 0.56 m/yr, or 50%. The vapor pressure of water decreases 45% when moving from 17°C [current global mean surface temperature (Harte, 1988)] to 8°C (CSG Network, 2013). This good agreement allows extrapolation to ~3/4 reduction in precipitation for the 20°C crisis. Agriculture constitutes approximately 67% of global water withdrawal (Alcamo et al., 2010) and 75% of human water consumption (evaporation) [not including rain-fed agriculture and silviculture (tree growing)] (Wallace, 2000). Agriculture in the 10°C crisis case would be limited at most and it would be nonexistent in the 20°C crisis case. Desalination currently supplies ~5 Gt/yr (Millennium Ecosystem Assessment, 2005). Given that irrigation is 1200 Gt/yr consumption (Millennium Ecosystem Assessment, 2005), the current non-rain-fed consumption is ~1600 Gt/yr, desalination is only ~0.3%; therefore, this could not make an important impact in the short term. Also, desalination is energy intensive.

Of course humanity would still need to produce food, but it would require far less water than plants. This is because plants have considerable evaporation loss in the process of extracting carbon from the low concentration carbon dioxide in the atmosphere. The equivalent relative humidity is low, even in the tropics, because of solar heating of the leaves. Hence, 1 ton of grain requires ~2000 tons of water (Hoekstra and Chapagain, 2005). Furthermore, water consumption is much higher now because of the large food losses, which we previously discussed, and would be significantly reduced during a crisis.

Fishing and food from leaves would require negligible water. In addition, the relative humidity in the tropics should remain high, allowing mushroom and bacteria growth, especially because of precipitation, the initial excess water in the logs, and water production from oxidation of the logs. Effective industrially produced food as discussed in the last Chapter requires a water amount about one order of magnitude larger than the dry weight of the food produced. The critical case is the 20°C crisis, and we assume conservatively only one-fourth as much consumption available (ignoring the reduced plant-cover transpiration). Beef production is incredibly water intensive. The drinking water for cattle is a factor of 110 greater than the wet weight of the meat (Beckett, 1993), and we assume that the meat makes up half of the edible calories. Because of one offspring per cow, the water consumption of the mother ("mother overhead") roughly doubles the consumption per calorie produced (Byerly, 1967). Ruminants would not become a full solution for several years, so there would be time to ramp up water supply. Therefore, the critical case would be rats or chickens. For these animals, the mother overhead is quite small, roughly cutting the water use in half. Therefore, the water consumption would be approximately 40 Gt/yr. This is roughly 10% of the available 20°C crisis consumption. Since eliminating irrigation would reduce noncrisis consumption to the available consumption, this means that the nonirrigation consumption must be reduced by 10% as well. Furthermore, recycling of the excrement, especially the urine, would reduce the net drinking water requirement for animals. In the particular case of cellulose-digesting beetles, since they are cold-blooded and consume their own excrement, water requirements should be negligible.

The greatest water consumption among the alternative food sources could be indoor growing of bacteria and mushrooms. Since the bacteria could convert at higher efficiency and a higher rate, mushrooms would be the critical case. With the high density of mushroom trays in buildings, the relative humidity would become high. This reduces the water lost to evaporation. Furthermore, the oxygen diffusing in and carbon dioxide diffusing out occur at much greater gradients than carbon dioxide diffusing into plants. Therefore, the water consumption of mushrooms would be orders of magnitude smaller than plants. Furthermore, the fact that there will be fewer plants in the natural ecosystems would increase runoff as a fraction of precipitation (drainage into the soil eventually becomes runoff). Therefore, the global supply of water would be adequate.

However, the regional picture is more challenging. Ironically, the dry areas with extensive irrigation systems would have excess water because little irrigation would be required. It is the areas without irrigation systems where water supply would be critical. Society could transport water to these areas, but this would be energy intensive and would require extensive transportation infrastructure. Humanity could relocate some water consumption to water-rich areas, but conservation in the electric power, industrial, and municipal sectors will likely be the primary solution.

Simple cycle natural gas turbines require no cooling water, and combined cycle natural gas turbines require only about one half as much water as conventional thermal power plants and a third of nuclear plants (Kehlhofer et al., 2009). Therefore, water use in this sector could be dramatically reduced by favoring gas turbine generators. Another way of reducing cooling water requirements for power plants and buildings is to repurpose air conditioning heat exchangers. With the ambient lower temperatures and greater temperature range tolerated by the people during a crisis, this would require considerably fewer air conditioners. These heat exchangers typically transfer heat from refrigerant to air, but it would be straightforward to change to water that currently goes to cooling towers, and retaining air on the other side of the heat exchanger.

Industrial production of nonessential goods (of which there are a lot when considering pure survival) would already be eliminated in order to conserve energy, and this has the added benefit of reducing water consumption. Furthermore, industry could implement some retrofits quickly, such as changing from once-through systems to multiple pass.

There are many options to reduce municipal water use. First, people could eliminate lawn irrigation. People could capture rainwater and reuse grey water (e.g. flushing toilets with sink drain water). People could displace some of the capacity in flush toilets, and not flush every use. People could use low-flow shower heads, turn off the water for lathering, and reduce showering frequency. People could also reduce clothing washing frequency, make the loads fuller, and use neighbors' low-water use washers. In addition, people could use low-flow faucets, turn off water when not in use, and fix many leaks. Actual human drinking water requirements are negligible compared to other uses. Unsustainable withdrawal of surface and groundwater could make up any shortfall, which interestingly is the current practice in many locations. Relocating animals and their drinking requirements is one way of taking advantage of these sources of water. Therefore, with unsustainable withdrawal and the considerable quick conservation potential in the use of water by electric power, industry, and municipalities, even regional water supply appears feasible.

An additional issue is the water distribution pipes freezing due to low air temperature. Possible solutions include digging up the pipes and burying them deeper, heating the water at the water treatment plant and at people's houses for sewage, insulating the ground above the pipes, piling soil above the pipes and worst case letting them run slowly continuously. All of the solutions involve an investment in energy and the last a significant investment in water as well (though this is not a consumption of the water, so it can be reused for other purposes). In some areas where the energy needed to keep pipes from freezing becomes prohibitively large, relocation is an option to warmer environments.

An additional issue related to water is soil erosion. In the 20°C crisis, even though there will be significantly less precipitation, because plants could not live, there would be increased soil erosion in the areas that were not frozen. This would cause short-term problems while the sun is still blocked, and long-term problems for the agricultural recovery. One potential short-term problem is disruption of fisheries, but there would be negligible fisheries in the 20°C crisis already because of little light. Another short-term problem is drinking water contamination, and a solution is settling ponds. Categories of solutions to the long-term problems include reducing erosion, adapting to the remaining soil, and using alternate growing environments. Some methods of reducing erosion include adding ground cover (some will happen naturally from dead vegetation not turned into food) such as gravel, polymer film, and metal foil. One method of adapting to the remaining soil is adding amendments (much vegetation would be turned into food, but there would likely be significant peat available). Other ways of adapting to the remaining soil include transporting soil from areas that had thick soil layers and selecting plants that can tolerate the poor soil. Some techniques of constructing alternate growing environments include crushing rock to create soil, hydroponics (which does not require soil), CO_2-fertilized algae-growing pools, and the ocean/natural freshwater bodies.

8.4 NUTRITION AND TASTE

Following the solutions outlined in Chapter 7, if the majority of humanity's diet were almost exclusively animals and/or typical bacteria, there would be very little carbohydrates. Again, this could provoke the Atkins-diet response (in some regions with epidemic obesity this may not seem like a problem at first (U.S. CDC, 2014), but dieting is not sustainable over years for anyone). Therefore, supplementing with stored grain in non-worst-case scenarios, leaf extraction/chewing/tea, mushrooms, and/or industrially produced sugar would be important. The converse problem is that insufficient protein could occur if the diet is primarily chewing/tea and industrially produced sugar. Then the ruminants, other animals, and possibly bacteria would be important diet additions. A diet based on high-carbohydrate bacteria or leaf concentrate and reclaimed sugar could be fairly balanced in terms of macronutrients.

Mineral supplements would be easy to provide and having some variety in food sources would improve the vitamin content. Also, the recommended daily allowances of vitamins typically have a safety margin. However, in the worst-case scenario of having to provide all vitamins in the form of supplements, this is approximately 0.5 g per day according to the United States recommended daily allowance (Otten et al., 2006). Since a small fraction of the global population currently takes full supplements, it is unlikely that industrial capacity for producing these vitamins could ramp up in time. However, engineered micro-organisms that produce each of these vitamins may scale up very quickly and consume a small amount of feedstocks, but this is for future work.

Although as engineers we cannot help but focus primarily on technical feasibility, here we acknowledge cultural limitations to our approach and provide a brief note to address mechanisms to ensure the taste and "mouth feel" acceptability of alternative foods. In the 2013 movie, Snowpiercer, a failed global-warming experiment kills off most life on Earth except for a few that boarded the "Snowpiercer", a self-sufficient train where a class system evolves. Curtis, the "hero" in the back of train, at one point explained how when he and the other low-class inhabitants first got onto the train there were no protein blocks (or food of any kind) and they were forced to cannibalize each other to survive. He admits killing and eating the mother of his second in command and that he is ashamed to know that "babies tasted best." Yet, when he discovers that the protein blocks later provided by the head of the train (upper class), which they had been eating for years, were made from insects he was disgusted. Apparently even fictional cannibals find eating insects disgusting. There is a strong cultural bias against eating insects in the United States and some other global cultures (although many cultures throughout the world routinely eat insects). For example, a commenter on Snowpiercer said "Insect diet, huh? I vote non-sustainable extinction." (Meat Trademark, 2014). That said, we believe hunger would likely overwhelm humanity's current arbitrary selective tastes given a serious crisis. Yes, even the beetles will look good to you when you are starving. In addition, humanity will still have the access to the discipline now known as food science that has brought us an awe-inspiring selection of processed foods in the grocery store. The same techniques that can make soy beans taste like turkey (tofurkey), ground beef (textured vegetable protein (TVP)), chocolate milk (Silk) or bacon (veggie bacon strips) can be used to make ground beetle and bacteria match the taste and mouth feel characteristics pleasing to the globe's varied taste buds.

We are hardly the first to suggest eating insects would be more sustainable (Oonincx et al., 2010) and even the UN is now urging people to eat insects to fight world hunger (BBC, 2013). About 2 billion people already purposefully make insects part of their diet (see Figure 8.2) (BBC, 2013). There are extensive DIY sites on growing mealworms (usually to feed to chickens) (e.g., Violett, 2012) and Tiny Farms recently developed a platform for edible insect farming, created an open-source mealworm farm shown in Figure 8.3 (Tiny Farms, 2014). Such farms could be laterally scaled now to provide humanity with an inexpensive and efficient source of protein and could be used to improve the overall efficiency in a post-disaster food system. In a disaster scenario and perhaps even in non-crisis times, a plate like that shown in Figure 8.4 of slightly spiced and cooked mealworms, will actually grow to be appetizing for Americans and other cultures that do not currently consume insects purposefully..

As well, there is already considerable storage of many spices at the household level. Also, the likely drying method is by burning local

Fig. 8.2. Deep-fried insects for human consumption sold at food stall in Bangkok, Thailand (photo credit: Takoradee, CC-BY-SA).

Fig. 8.3. Initial beta version of the Open Bug Farm Mealworm Farm Kit (Tiny Farms, 2014).

Fig. 8.4. A plate of mealworms. Photo courtesy of Rob Loftis (CC-BY).

biomass, which would provide smoke flavor. Furthermore, industry and bacteria produce some food flavorings, which could continue. In addition, salt will remain inexpensive, and at least some sugar could be produced through industrial digestion. As a last resort, as mentioned above, artificial light could be used to produce spices.

8.5 BIODIVERSITY

Many would argue that we should not only save humans but the rest of the Earth's inhabitants from extinction during a catastrophe. What would a modern day Noah's ark really need in terms of food? The caloric requirements of preserving 100 individuals of every species would actually be fairly modest. For instance, 5,000 mammal species with an average individual metabolic requirement an order of magnitude greater than a human (e.g. saving the whales) would be equivalent to 5 million humans, more than three orders of magnitude less than keeping all the people alive. There are a tremendous number of insect species, but their individual caloric requirements are very small.

However, there are other constraints. Many animals are specialized feeders. It would be fairly straightforward to keep the carnivores alive because there would be stored meat and many of the food production options are high-protein. For specialist herbivores, humanity could preserve their food, such as bamboo for pandas. However, if this were not possible, artificial light production of plants would make preserving these species very expensive. Housing the large whales would be very difficult, though it may be possible to feed them in the ocean. For insects, scientists have identified only a small fraction of the total species, let alone being able to capture them and keep them alive. As a last resort, humans could preserve individuals or DNA samples, such as with cryogenic preservation. There is already extensive work on this mechanism in the Frozen Ark Project, which aims to save samples of frozen cells containing DNA from endangered animals before they go extinct. This project would need to be expanded to cover all of the many species on the Earth. It would be very expensive to keep plants alive, but collecting seeds would generally preserve them. Fungi, protists, bacteria, and archaea (an early form of life) would probably be just fine on their own and not need human support.

8.6 OTHER PROBLEMS

This section covers other problems associated with the solutions provided in Chapter 7. All of them are solvable.

If you are reading this book in the bathroom, you may have already wondered about the following question: If we are chopping down all the trees for food what about toilet paper? There would be plentiful wood feedstock in the form of dead trees outside the tropics that would be more difficult to use for food production. However, much of the forestry equipment may be utilized in the tropics for food production. Thus, it may be necessary to limit dramatically global consumption of lumber, paper, and paperboard, and this would be technically feasible. Solutions include substitution, recycling, and mining landfills. Other fibers, such as cotton, hemp, jute, wool and silk, would be in very short supply, but would have similar solutions.

Generally, given the crises in Chapter 3 there will be a plume of dust, toxic gases, high temperature, and/or ionizing radiation emanating from the source(s). The dry deposition of the larger particles would be relatively fast. Then the small particles and soluble gases would be rained out in a few days. High temperature would also decay relatively quickly due to radiation upward and downward. An exception could be if an impact occurred on land and threw hot rocks around the world, causing widespread fires. However, the intensity of the ejecta decays with the third power of the distance (Collins et al., 2005), so it too would still generally be localized. Of course dust, gases, and ionizing radiation particles could be injected into the stratosphere, and therefore have a global impact. However, the fallout rate is much lower because it would happen over years, so the intensity would be orders of magnitude lower than the local impacts. The only exception for considerable global exposure with large disasters would be low solubility gases, such as carbon monoxide. Therefore, the effect on food production of these factors in most cases would be relatively small in the global sense.

8.7 COOPERATION: THE ELEPHANT IN THE ROOM

Although it is technically viable to preserve all of human life and a good selection of other species given even the absolute worst global catastrophes outlined earlier, none of these solutions will work for all of humanity without global cooperation. We are not naïve optimists sipping lattes on our private yachts. We understand the magnitude of the problem. Although, technically nearly trivial, practically this is an enormous barrier as humans historically and currently have been savage to one another (for references read any history book or turn on any news program any time). The news can be somewhat misleading as there is some room for cautious optimism as statistics show a steady decline in human violence (Pinker, 2011).

Despite this, there are two sub-problems to solve. First, there is the issue of the poor of the world not being able to afford high-priced food, with possible solutions of subsidies, loans and charity. In addition, there will be an enormous need for new work (much of it will be manual labor – e.g., cutting down dead trees, harvesting food in a fundamentally new way, retrofitting industrial equipment, etc.) that could provide food-ration employment for those in the developing world able to work (and food would also go to their families). Second, there is the issue of maintaining global cooperation because trade would be very important. Although it will be in the vast majority of humanity's interest to cooperate, conflict is likely to be in some countries' best interests, so this will be the most challenging factor to overcome. Even if outright conflict were not to occur, there would be a strong temptation to hoard food at the personal, family, community, organizational, and national levels. The optimist would hope that when faced with a common enemy that threatens the survival of the species (and many other species on Earth) humanity would pull together and collaborate for the common good to get everyone through. If that does not work, some concrete steps for solving these problems are outlined in Chapter 10.

REFERENCES

Alcamo, J., Doll, P., Henrichs, T., Kaspar, F., Lehner, B., Rosch, T., Siebert, S., 2010. Global estimates of water withdrawals and availability under current and future. 'business as usual' conditions Hydrological Sciences Journal 48, 339–348.

BBC, 2013. UN urges people to eat insects to fight world hunger. http://www.bbc.com/news/world-22508439.

Beckett, JL, Oltjen, JW, 1993. Estimation of the water requirement for beef production in the United States. Journal of Animal Science, 71:818–826.

Branker, K., Pathak, M.J.M., Pearce, J.M., 2011. A review of solar photovoltaic levelized cost of electricity. Renew. Sustain. Energy Rev 15 (9), 4470–4482.

Byerly, T.C., 1967. Efficiency of feed conversion. Science 157, 890–895.

Collins, G.S., Melosh, H.J., Marcus, R.A., 2005. Earth impact effects program: A Web-based computer program for calculating the regional environmental consequences of a meteoroid impact on Earth. Meteorit. Planet. Sci 40, 817–840.

CSG network. 2013 Vapor pressure calculator. CSG network. Available: http://www.csgnetwork.com/vaporpressurecalc.html.

Harris, A., 2011. Don't wait for the blackout. Eng. Technol 6, 60–62.

Harte, J., 1988. Consider a Spherical Cow: A Course in Environmental Problem Solving. Sausalito, CA, University Science Books.

Hoekstra, A.Y., Chapagain, A.K., 2005. Water footprints of nations: Water use by people as a function of their consumption pattern. Water Resour. Manag 21, 35–48.

IEA, 2012. World Energy Outlook 2012. International Energy Agency.

Kammen, D.M., 2006. The rise of renewable energy. Sci. Am 294, 84–93.

Kenny, R., Law, C., Pearce, J.M., 2010. Towards real energy economics: Energy policy driven by life-cycle carbon emission. Energy Pol 38, 1969–1978.

Kehlhofer, R., Rukes, B., Hannemann, F., Stimimann, F., 2009. Environmental Considerations. Combined-cycle gas and steam turbine power plants. 3rd edition PennWell, United States, pp. 261–275.

Legates, D., Willmott, C., 1990. Mean seasonal and spatial variability in gauge-corrected. global precipitation, International Journal of Climatology, 10, 111–127.

Meat Trademark http://movies.stackexchange.com/questions/18708/what-is-in-the-protein-bars-in-the-snowpiercer.

Millennium Ecosystem Assessment, 2005. Ecosystems and Human Well-Being: Current State and Trends. Island Press, United States.

Ooninex, D.G., van Itterbeeck, J., Heetkamp, M.J., van den Brand, H., van Loon, J.J., van Huis, A., 2010. An exploration on greenhouse gas and ammonia production by insect species suitable for animal or human consumption. PloS one 5 (12), e14445.

Otten, J.J., Hellwig, J.P., Meyers, L.D., 2006. Dietary reference intakes (DRIs), 560.

Pearce, J.M., Denkenberger, D., Zielonka, H., 2009. Accelerating applied sustainability by utilizing return on investment for energy conservation measures. Int. J. Energy, Environ. Eco 17 (1), 61.

Pinker, S., 2011. The better angels of our nature: Why violence has declined75Vikig, New York.

Sims, R.E., Rogner, H.H., Gregory, K., 2003. Carbon emission and mitigation cost comparisons between fossil fuel, nuclear and renewable energy resources for electricity generation. Energy Pol 31 (13), 1315–1326.

Tiny Farms. 2014. Open Bug Farm Mealworm Farm Kit Now Available! http://www.tiny-farms.com/blog.

U.S. CDC. Overweight and Obesity. Available http://www.cdc.gov/obesity/data/adult.html.

Department of Energy, U.S., 2011. Energy Efficiency and Renewable Energy, 30th edn Transportation Energy Data Book. U.S. Department of Energy, Springfield, VA.

U.S. Department of Transportation, 2001. National household transportation survey.

Wallace, J., 2000. Increasing agricultural water use efficiency to meet future food production. Agric. Ecosyst Environ 82, 105–119.

Violett, L., 2012. How to Grow Mealworms (includes step-by-step photos) http://home.earthlink.net/~lviolett/mealworms.html.

CHAPTER 9

Moral Hazard

9.1 MORAL HAZARD OF WRITING THIS BOOK

What if writing this book creates the possibility that awareness of a food backup plan will result in less effort to prevent these crises? There is a nonzero probability that we could have increased the chance for humanity's collapse rather than prevent it. This would be the exact opposite of our goal of making the survival of humanity more probable armed with knowledge and a beginning sketch of a plan for feeding everyone given one (or more) of the scenarios discussed in Chapter 3. Our intentions we hope are quite transparent and, as you are reading this, you know we decided the benefit outweighed the risk. To make that clear, we have done some basic calculations.

9.2 NUCLEAR STOCKPILES INCREASING IF REDUCED RISK OF NUCLEAR WINTER CAUSING MASS STARVATION?

Nuclear winter is the crisis over which humanity has the most technical control and, as we saw in Chapter 3, poses the most serious threat. Nuclear winter is only possible because humanity as a whole has far more nuclear weapons than would be necessary in any interhuman conflict. Yet, powerful countries with them cling to them like a child clings to a security blanket. First, it is not clear that possessing nuclear weapons makes a country safer (Schwartz, 1998; Paul, 2000). However, even if the United States and Russia wanted to retain nuclear weapons for security reasons, they could reduce the number by an order of magnitude and still have the power to inflict horrific economic and loss-of-life damages by destroying large city centers of any potential enemies, thus providing deterrence (Smith, 2013). If the two largest hoarders of nuclear weapons reduced their stockpiles to deterrence minimums, the risk of nuclear winter would be reduced dramatically globally for not only those country's citizens, but also the rest of the world. There are also other steps to reduce the risk of nuclear winter (Barrett et al., 2013). Mikhail Gorbachev

Feeding Everyone No Matter What. http://dx.doi.org/10.1016/B978-0-12-804447-6.00009-7

explicitly stated that a motivating factor for reducing the nuclear arsenal of the USSR was the studies predicting nuclear winter and therefore destruction outside of the target countries (Toon et al., 2008). However, despite the knowledge of the possibility of nuclear winter, the nuclear arsenals remain large enough to potentially cause nuclear winter. Shouldering this risk actually costs considerable money. The costs are so large and poorly tracked that putting an official price tag is not possible (U.S. GAO, 2005). However, just for the effort to upgrade and maintain the United States' 5,113 warheads, to replace old delivery systems and to renovate the aging facilities where nuclear work is performed, a study by the nonpartisan Stimson Center estimated costs would be at least $352 billion over the coming decade to operate and modernize (Priest, 2012). This is objectively a questionable use of resources. At the very least, each country with nuclear weapons should reduce their own numbers below the nuclear winter limit and then work with one another to ensure that the aggregate is below this value. It should be intuitively obvious that no one wins any form of nuclear war that results in a global catastrophe capable of wiping out humanity as a whole. It should be equally clear that if military or civilian leadership is so focused on maintaining and extending the power of their own bureaucracy that it blinds them to this fact at the expense of de-facto security for their nation, they should be removed from authority.

9.3 GREENHOUSE GAS EMISSIONS INCREASING IF STARVATION RISK OF ABRUPT CLIMATE CHANGE DIMINISHED?

The threat from nuclear war and the potential of nuclear winter are poignant and easy to understand. There is a clear and present threat of global warming (anthropogenic abrupt climate change) (Schwartz and Randall, 2003; Schneider, 2004), which is more abstract and harder for the public to understand and conceptualize. As both the threat and the causes of climate change are more dispersed and some of the most powerful companies in history are aligned to continue and accelerate fossil fuel combustion, climate change and global warming have been muddled (often intentionally) in the mind of the public [particularly in the U.S. (Leiserowitz et al., 2011)] (Hoggan et al., 2009; Hoffman, 2011; Dunlap and McCright, 2011; Washington and Cook, 2011; Bain et al., 2012).

Attempts by the scientific community to extract climate change from the U.S. culture wars have been largely ineffective (Leiserowitz et al., 2011; Bain et al., 2012) and little has been done to effectively prevent global climate change (Yone et al., 2007). It is possible that because this book presents relatively painless solutions to global hunger caused by abrupt climate change, there will be even less incentive to stop carbon emissions, putting humanity at greater risk. This again is ill-advised. However, it is also possible that the literal distastefulness of some of the food solutions in this book could give those aligned against climate change mitigation a pause for thought. Regardless, it is unlikely that food solutions will have any serious impact on the debate (Leiserowitz et al., 2011; Bain et al., 2012).

9.4 MORAL HAZARD OF OTHER RISKS

Furthermore, the backup plan presented here could reduce the damage associated catastrophes over which humanity has no or very little control (e.g. supervolcanos). The only cases for which moral hazard appears to be important are the super weed, super bacterium, super crop pest, and super crop pathogen. In all these cases knowing that these threats could be largely mitigated with a long list of solutions from Chapter 7 could make humanity less vigilant at preventing the risks from occurring. As we have discussed before, these threats are all relatively modest compared to the more intense sun-blocking scenarios. In addition, the scenarios to lead to these threats (e.g., bio-terrorists creating a super crop pathogen) are unlikely to be influenced by this book in any material way.

9.5 CONCLUSION: WHY YOU ARE ABLE TO READ THIS BOOK NOW

Therefore, despite the relatively small moral hazard dilemma, we believe humanity would be much better off with a viable backup plan. The further we as a species can develop the plan, the lower the probability of our own extinction. The steps necessary to do this are outlined in the final Chapter.

REFERENCES

Bain, P.G., Hornsey, M.J., Bongiorno, R., Jeffries, C., 2012. Promoting pro-environmental action in climate change deniers. Nat. Climate Change 2 (8), 600–603.

Barrett, A.M., Baum, S.D., Hostetler, K.R., 2013. Analyzing and reducing the risks of inadvertent nuclear war between the United States and Russia. Sci. Global Secur 21, 106–133.

Dunlap, R.E., McCright, A.M., 2011. Organized climate change denial. The Oxford Handbook of Climate Change and Society, 144–160.

Hoffman, A.J., 2011. Talking past each other? Cultural framing of skeptical and convinced logics in the climate change debate. Organ. Environ 24 (1), 3–33.

Hoggan, J., Littlemore, R.D., Ball, T., 2009. Climate Cover-Up: The Crusade to Deny Global Warming. Greystone books, Vancouver.

Leiserowitz, A., Maibach, E., Roser-Renouf, C., Smith, N., 2011. Global Warming's Six Americas. Yale Project on Climate Change Communication.

Paul, T.V., 2000. Power Versus Prudence: Why Nations Forgo Nuclear Weapons. McGill Queens University Publisher.

Priest, D., 2012. U.S. nuclear arsenal is ready for overhaul. Washington Post. September 15, 2012.

Schneider, S.H., 2004. Abrupt non-linear climate change, irreversibility and surprise. Global Environ. Change 14 (3), 245–258.

Schwartz, S.I., 1998. Atomic Audit: The Costs and Consequences of U.S. Nuclear Weapons Since 1940. Brookings Institution Press, Harrisonburg, VA.

Schwartz, Peter, Doug, Randall, 2003. An abrupt climate change scenario and its implications for United States national security. CALIFORNIA INST OF TECHNOLOGY PASADENA CA JET PROPULSION LAB.

Smith, R.J., 2013. Obama administration embraces major new nuclear weapons cut. The Center for Public Integrity. http://www.publicintegrity.org/2013/02/08/12156/obama-administration-embraces-major-new-nuclear-weapons-cut

Toon, O., Robock, A., Turco, R., 2008. Environmental consequences of nuclear war. Phys. Today 61, 37–42.

U.S. G.A.O., Aug 4, 2005. Actions Needed by DOD to More Clearly Identify New Triad Spending and Develop a Long-term Investment Approach. GAO-05-962R: Published.

Washington, H., Cook, J., 2011. Climate Change Denial: Heads in the Sand. Earthscan.

Yone, G.W., Lasco, R.D., Ahmad, Q.K., Arnell, N.W., Cohen, S.J., Hope, C., Janetos, A.C., Perez, R.T., 2007. Perspectives on climate change and sustainability. Change 25 (48), 49.

CHAPTER 10

Serious Prepping: A Guide to Necessary Research

10.1 POLICY IMPLICATIONS

The analysis in this book has focused primarily on the technical feasibility of complete human food supply in extreme circumstances. However, there are several policy implications during both normal times and disasters, to prepare to ensure the bulk of humanity survives in even the most challenging of situations as discussed previously.

Despite the enormous global catastrophic risks summarized in the first three Chapters, this book has shown that it is technically viable to maintain the entire human population with existing vegetation even in extreme global catastrophes. Unfortunately, even in these peaceful times with minor challenges by comparison (e.g., agriculture works, plenty of sun, water, area, etc.), humanity fails to feed everyone. Under nearly all moral and ethical systems, this is unacceptable and those with either political or economic power should be simply embarrassed. They should be ashamed that they are not powerful enough to provide even simple food for their populations. This shame should extend from lowly village mayors to the regional governors, to the most powerful leaders of the free world who could take it upon themselves to end human starvation under their watch. How powerful can a leader be if his/her people are starving? How dominant of a species are we if thousands still starve to death every day? It is long past time we started thinking of a humanity as a whole and cared for our own. Our leaders are not currently doing this and the numbers make it obvious: 870 million people do not have enough to eat (United Nations Children's Fund, 2006) and under-nutrition contributes to over 6 million deaths of children under 5 each year in developing countries.

If we assume that charitable food aid will always lag demand, we can still "teach a person to fish" rather than provide them with the fish.

Feeding Everyone No Matter What. http://dx.doi.org/10.1016/B978-0-12-804447-6.00010-3

There are well-known permaculture methods (Mollison, 1988; Mollison and Slay, 1991), agricultural improvements for sustainable development (Reijntjes et al., 1992; Pretty, 1995; Roling and Wagemakers, 2000; Tilman et al., 2002) and open-source appropriate technologies (OSAT)[1] that we could employ to feed everyone (Pearce et al., 2008; Pearce and Mushtaq, 2009; Pearce et al., 2010; Zelenika and Pearce, 2011; Pearce, 2012; Pearce et al., 2012; Sianipar et al., 2013). These ideas and methods that work must be spread throughout the world more rapidly than we have been able to do historically. With the growth and success of the Internet, there is now no technical reason that the best solutions for providing for food needs should not be made freely available to everyone that needs them. The OSAT community provides a clear model to build upon.

Open source is a development method borrowed from the enormous success of the open-source software industry (Weber, 2004). It has been applied in many fields from the creation of sophisticated scientific equipment (Pearce, 2014) to 3D printers (Jones et al., 2011) to simple appropriate technologies for sustainable development (Hazeltine and Bull, 1998; Pearce, 2012). These OSATs harnesses the power of distributed peer review and sharing to solve problems in the developing world. Appropedia.org is an example of the development of open-source appropriate technology (see Info Box 10.1). Appropedia is an open-wiki that anyone can edit to find collaborative solutions for sustainability, poverty reduction, and international development, with a particular focus on appropriate technology. There anyone can learn both how to make and use AT free of concerns about patents. At the same time, anyone can also add to the collective open-source knowledge base by contributing ideas, observations, experimental data, deployment logs, etc. The built-in continuous peer-review can result in better quality, higher reliability, and more flexibility than conventional design/patenting of technologies. The free nature of the knowledge also obviously provides lower costs, particularly for those technologies that do not benefit to a large degree from large scale of manufacture. Finally, OSAT prevents

---

[1] Appropriate technologies can be both high-tech like solar photovoltaic cells or low-tech like a simple rope pump. What AT has in common is that it encompasses technological choice to be appropriate for a specific location. In general, AT is small-scale, decentralized, energy-efficient, environmentally sound, and locally controlled (Hazeltine and Bull, 1998).

intellectual property restrictions from preventing access to useful tools. This is particularly important in the context of technology focused on relieving suffering and saving lives in the developing world (Pearce and Mushtaq, 2009). In the next section, we will discuss how we can borrow OSAT methods and tools to begin to crowd source some of the research needed for the solutions proposed in this book. After an initial surge in the 1970s, AT development declined in favor of large developmental projects like dams. It has begun resurgence with the OSAT methods on the Internet, but is still far from ubiquitous (Zelenika and Pearce, 2011). At the national and international levels (e.g. the UN) a commitment to making research and technology development open source or placed in the public domain could go a long way to stopping hunger worldwide now and prepare for even the worst possible futures at the same time. The cases for open access of scholarship, in general, and OSAT, in particular, have been forcefully made elsewhere (Willinsky, 2006; Pearce, 2012). The general premise is that the greatest benefit will come from the research if it is spread widely and that this should be done so, particularly in the cases where the public is funding the research.

Thus, food policies can prioritize the research outlined in this Chapter to improve significantly the technical viability of all the solutions analyzed here for disaster scenarios. At the international level, policy could address the most challenging issue of the poor of the world not being able to afford high-priced food, such as with subsidies and loans or forms of training and economic development. Finally, and most importantly, policies are needed to pre-emptively ensure and maintain global cooperation in the event of agriculture-disruptive catastrophes because trade would be very important to ensure global human survival.

10.2 APPLYING SOLUTIONS TO CATASTROPHES NOW TO PROVIDE FOOD FOR TODAY'S HUNGRY

In the regions where hunger-related death is most common and the poor do not have access to conventional agricultural resources, humanity could utilize some of the solutions outlined here directly. For example, mushrooms can grow on waste products, including forest residue. These applications should be tested and then scaled and deployed where they are needed and appropriate. As mushrooms can grow on waste products, people should recognize them as an environmentally friendly food source. This could increase their consumption, reducing land and water use. There is currently a very large supply of materials that mushrooms could utilize (but many cellulose digesters cannot) in the form of silviculture (logging) residues. For cellulose digesters, the practice of feeding excrement from other animals could be expanded. With this high-nitrogen food source, lower nitrogen sources could be used in addition, such as agricultural residues that are not green and tree leaves that have been depleted of their nutrients (and shed). These practices have the potential of dramatically reducing the ecological footprint associated with cellulose digesters, which form a considerable fraction of the food and overall ecological footprints. Also, extracting human food from leaves and other agricultural residues would reduce environmental impact and may provide more healthy food for locations that need it.

Macronutrient (nitrogen and phosphorus) fertilization of the ocean could also produce a considerable amount of food. This would take pressure off terrestrial ecosystems, preserving biodiversity. By preventing forest destruction, CO_2 release would be reduced thus helping contain climate change. Furthermore, some of the carbon produced in the ocean would sink to the ocean floor, sequestering CO_2, which would again help in a small way with the climate destabilization under way. Thus, policies to support deployment of the macronutrient fertilization of the ocean could increase food supplies substantially for some communities, while also playing a role in global warming reductions. Moreover, policies could promote the dissemination of clear instructions on extracting human food from leaves and other agricultural residues to those most in need. Similarly this would apply to other techniques and OSAT.

Economists may be forgiven for questioning additional food production solutions proposed here as being viable for reducing global hunger now, as it is the inability to produce food at exceptionally low costs that leads to starvation today. However, these solutions may result in lower cost food overall and many can be implemented by individuals in marginal economic situations allowing for lateral scaling.

10.3 FUTURE WORK

This analysis has focused on technical feasibility of food supply; however, future work is needed to improve existing technologies, to test solutions, and to investigate the economics and politics of making a backup food supply available for humanity. In addition, the following experiments broken into food, fiber, and supply problems are technical in nature and will reduce the cost and improve the viability of these alternative sources of food for global catastrophes and thus reduce the probability of conflict.

Food Growing Experiments include:
- constructing enclosures and racks for drying food, possibly with 3D printers so that the open-source digital designs can be quickly deployed by lateral scaling;
- growing photosynthetic crops in the crisis conditions of the tropics;
- determining realistic ramp rates, conversion efficiencies, and conversion rates of all the promising food options from Chapter 7.

Fiber Supply Experiments include:
- putting fresh wood and partially digested wood in rock crushers;
- putting partially digested wood in wood grinders;
- providing open source hardware designs for modifying a loader to grab or scoop logs, and modifying a heavy-duty truck to carry logs and loading the logs in different ways. Methods provided by Open Source Ecology and their Global Village Construction Set could be used as a model (OSE, 2014).

Future work should estimate labor for all of these with realistic workers. However, generally the solutions can utilize mechanical means (e.g., chainsaws, not manual saws); therefore, we do not expect labor to limit the technical feasibility.

Nutrition Experiments include:
- Humans eating pasteurized nearly completely decomposed: (1) wood and (2) nonwoody biomass.
- Humans eating most of each of the following: mushrooms, beetles, rats, rabbits, ruminants, sugar, fish, methane bacteria, leaf extract, and chicken.

- Future work also includes making sure there are bacteria that produce large amounts of each vitamin required for humans. Micronutrient supplements for food animals would be more difficult than for humans, so this requires further work.
- Another important experiment is modifying an existing industrial process to digest wood.

Other Experiments and Modeling

Future climate modeling work could investigate scenarios with nearly complete blocking of sunlight globally. Also, reducing the uncertainty in the regional impacts would be valuable. The interaction of food crises is generally less likely, but would be more severe, so it is important to study in the future. Engineering high-carbohydrate bacteria is important. It would also be useful to analyze the net impact of reducing waste now. Reducing the waste of food, water, and energy and increasing renewable energy would mollify slow climate change and possibly abrupt climate change. However, these actions would make it more difficult to respond to the other crises considered in this book. Projecting future capability to handle crises would also be valuable.

There is great urgency to perform follow-up research that increases the probability of success of these solutions. For instance, if a research project increased the probability of most people surviving versus most people dying by only 0.1% given a crisis, every day of delay of completion of the project would cost an order of magnitude 10 lives. The total probability of one of the high intensity disasters (nuclear winter, super volcano, or asteroid or comet impact) occurring in the next year is 0.1011% (showing non-significant figures for clarity). Without the solutions proposed in this book, given ~1 year supply of food, ~1/5 of people could survive for 5 years in the technical scenario. With a current population of roughly 7 billion, this would result in approximately 6 billion people dying. In the technical scenario with the solutions in this book, nearly everyone could be saved. If the future research increases the probability of success of these solutions by 0.1%, the expected value of the research per year is approximately 6000 lives. This means every day delay of the research costs an expected ~20 lives (order of magnitude 10). And of course when considering future generations, reducing the chance of losing civilization or our species, a tiny amount would mean many more expected lives saved.

10.4 PREPARATION

As was already mentioned, storing up food is a solution that requires considerable preparation and is not economically justifiable when there are other methods available to us. In the same vein, larger stocks of particular animals could be intentionally kept to accelerate the time at which these food sources would attain full human food supply. Another solution that requires significant preparation is genetically engineering crops to handle lower light and temperature. Another is building checks into the food system to prevent the spread of super organisms. A further solution is lab-grown meat using feedstocks of algae. Although it should be noted that this is currently a very costly technology. As the technology improves and scales, the costs would be expected to decline by orders of magnitude.

Another solution that requires preparation is producing a considerable fraction of human energy from cellulosic acid-conversion biofuels, which would allow the interruption of the process at the sugar stage. This could provide all human food, assuming that the harvest of sufficient dead biomass that is compatible with the conversion process was feasible. Alternatively, industry could precondition less-compatible biomass. Scaling up this solution in non-crisis time could be justified on the bases of replacing finite petroleum and reducing slow climate change, though it has the disadvantages of high water and land use if it does not use residue feedstocks.

Considering the ozone problem, nanoparticles injected into the stratosphere that primarily scatter UV radiation may be able to solve the ozone problem and not reduce light for plant growth below. Another possible solution is injecting ozone into the stratosphere. Work on these solutions could be justified even for preventing slow climate change.

In addition, basic research investigating the viability of the proposed mechanisms discussed in this book would considerably help preparation and human resiliency for any food-related crisis.

10.5 HOW CAN I PREPARE MYSELF AND MY FAMILY?

Of the promising solutions for global food supply, we do some calculations on the ones that we think would be feasible at the household level. These would require the storage of biomass or having a continuing

Table 10.1 Tons of biomass or fossil fuel required to feed a household of four for 5 years.	
Solution	Tons of Biomass/ Fossil Fuel Required
Sugar from enzymes	13
Bacteria eaten with fiber	40
Leaf extraction then mushrooms	43
Methane-digesting bacteria	17
Chickens	114
Rats	114
Beetles	61
Rabbits	16

supply of natural gas. They would also require having the organisms for food conversion. This would be easy for things that double very quickly, like bacteria and mushrooms. It would be moderate difficulty for things like chickens, rats, and rabbits. It would be more difficult for large animals like cows (unless you happen to live on a cattle farm), so we do not consider this further. The fish solution only works for the 10°C crisis and requires lots of water area, so this would not generally be feasible at the household level. Table 10.1 shows the number of tons of biomass or fossil fuel required to provide food for a family of four for 5 years (about 4 tons of dry food). The biomass is more weight than just storing food, but the biomass can easily be stored outside under a tarp (so it does not get wet). Also, the biomass would be cheap or even free (e.g., the wood chips left by utility crews along the road). A bioreactor for growing methane-digesting bacteria sounds complicated and expensive, but a compost bin is an example of a bioreactor with solid material. When using methane solutions, there is the complication of the possibility of natural gas leaks, so people should have access to a natural gas detector.

Families may also need some nutritional supplements for their animals and some fertilizer. Individuals will need to avoid inbreeding in their animals, and solutions include trading with another household, freezing sperm, and having more than 50 animals to start with. For families that do not have too much space, a good option could be keeping some rabbits and having some bacteria that are good at producing enzymes that break cellulose into sugar. Following conventional prepper or Mormon recommendations, it would be good to have some stored food while ramping up alternate food supply. Individuals, however, can do far more than help their own families.

10.6 WHAT CAN I DO?

If you are a person of good will, no matter what your skill set and education, there is much that needs to be done and that you can do to help feed everyone no matter what.

First, try to encourage international goodwill and collaboration whenever possible. Visit other countries and get to know their people, converse and collaborate with them online, invite foreigners to visit you, house exchange students, donate to charities that allow communication with sponsored children or families, help your children find pen pals abroad, help new immigrants integrate into your country while learning about their old one. Sign petitions online, vote, lobby, write to the newspapers, blogs and your representatives to encourage diplomatic and peaceful solutions to international problems, reduce and eliminate nuclear weapons and fund a Department of Peace.

Second, spread free and open source information that can help others as much as possible. Contribute to international sites like Appropedia, which catalogue solutions by learning how to edit a wiki (see Info Box 10.1), and then : (i) create and edit articles on relevant topics you know about or have read about in other sources (particularly closed sources not available to the world's poor), (ii) copy edit and improve the articles and add hyperlinks to make them more easy to navigate, (iii) if you know multiple languages translate the most useful articles to spread the ideas further, (iv) consider building and trying to improve OSAT that you may use yourself (e.g., drip irrigation for your garden is also extremely useful for small farmers in the developing world), and (v) solicit information from experts and encourage them to share their knowledge with the world more directly.

Third, consider becoming a citizen scientist, experimenting and publishing your results openly on the web. Take a good hard look at the science and engineering that needs to be done in section 10.3. Can you do any of it yourself? For example, figuring out how to grow rabbits on collected biomass might be a fun family activity and within most family

Information Box 10.1

How to Edit Appropedia.org

Appropedia is a wiki that anyone can edit, and thousands of people have already added to it to benefit the world. Figure 10.1 shows a screen shot of the Food and Agricultural Portal, which hosts hundreds of articles relevant to this book. Navigation is on the left of every page and the "Help" (blue arrow) will provide you with all the basics of wiki markup, which is just a formatting syntax that takes about 15 min to learn (and is the same as is used in Wikipedia). For example, placing double brackets around a word like [[this]] creates a hyperlink to the word on the page you are editing to a page of the same name. All key concepts and words on Appropedia should be hyperlinked – and this is an easy thing for new comers to do that helps improve the resource while they are learning.

  You can join Appropedia for free by clicking on "create account" (purple arrow) in the upper right hand corner of any page. The mini-form will take you less than 1 min to complete and prevent you from having to solve CAPTCHAs every time you want to edit. You can edit all the pages of Appropedia by clicking on the "edit" tab at the top of the page or on any of the "edit" hyperlinks in the subsection of the pages (circled in red). You can start a new page by searching for it (yellow arrow). If it does not already exist, the result will be a new red hyperlink – click on it and begin typing. When you are done click save and you have made the page. As soon as you have made a new page, people from all over the world can help make it better by editing it, just as you can with their pages. Do not worry about messing up, all changes are recorded in the history tab and can easily be fixed by you, subject matter experts or Appropedia's moderators and administration. Millions of people use Appropedia from all over the world (with the notable exception of North Korea). Join the community to learn and help one another.

budgets. Try experiments and then ensure your data are collected and merged with others. Are you good at cooking? If so see if you can make a delicious meal with beetles as the main ingredient. If your doctor allows it and you are up to it, consider running some of the extreme food diets on yourself and monitor your body's response. Combine alternate food sources such that you at least get some carbohydrate, protein, and fat, and make sure to take a multivitamin. A controlled trial of people eating leaf extract only would be both difficult and expensive to run through the normal scientific funding bodies (e.g. graduate student salaries would probably have to be raised). On the other hand, a crowd-sourced experiment, where people gather and process leaves for food for a month could provide extremely useful data and actually save the participants money (cut their food budgets to near zero). It should go without saying that you should obtain your physician's blessing, be very careful about self-experimentation and ensure that you are not eating poisonous or otherwise harmful organisms by accident.

  You can also contribute by donating money. Charities working on reducing global catastrophic risk and our vulnerability to it include:

- Global Catastrophic Risk Institute: Geographically distributed in the U.S. http://gcrinstitute.org/

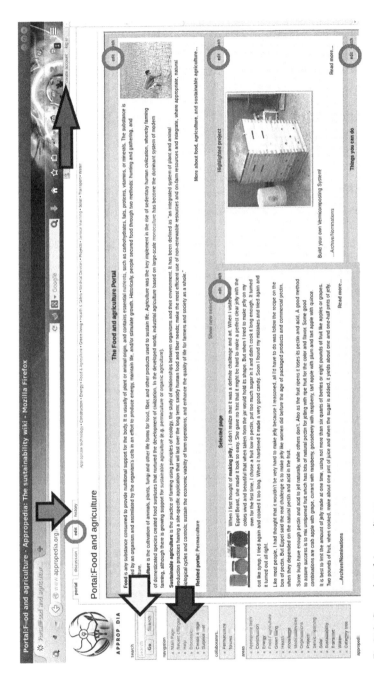

Fig. 10.1.   Screen shot of Appropedia.org

- Future of Humanity Institute: University of Oxford, U.K. http://www.fhi.ox.ac.uk/
- Centre for the Study of Existential Risk: University of Cambridge, U.K. http://cser.org/
- Center for Risk Studies: University of Cambridge, U.K. http://www.risk.jbs.cam.ac.uk/

Most importantly, share what you have learned with others so we can feed everyone no matter what.

REFERENCES

Hazeltine, B., Bull, C., 1998. Appropriate Technology; Tools, Choices, and Implications. Academic Press, Inc.

Jones, R., Haufe, P., Sells, E., Iravani, P., Olliver, V., Palmer, C., Bowyer, A., 2011. RepRap–the replicating rapid prototyper. Robotica 29 (01), 177–191.

Mollison, B., 1988. Permaculture: a designer's manual. (Tagari Publications: Tasmania, Australia).

Mollison, B., Slay, R. M., 1991. Introduction to Permaculture. Tagari Publications, Tyalgum, Australia.

Open Source Ecology. 2014. Machines: Global Village Construction Set http://opensourceecology.org/gvcs/

Pearce, J.M., 2012. The case for open source appropriate technology. Environment, Development and Sustainability 14 (3), 425–431.

Pearce, J.M., 2014. Open-Source Lab: How to Build Your Own Hardware and Reduce Research Costs. Elsevier.

Pearce, J.M., Mushtaq, U., 2009. September. Overcoming technical constraints for obtaining sustainable development with open source appropriate technology. Science and technology for humanity (TIC-STH), 2009 IEEE Toronto International Conference, 814–820, IEEE.

Pearce, J., Albritton, S., Grant, G., Steed, G., Zelenika, I., 2012. A new model for enabling innovation in appropriate technology for sustainable development. Sustainabil. Sci., Pract. Pol 8 (2), 1012–1067.

Pearce, J.M., Blair, C.M., Laciak, K.J., Andrews, R., Nosrat, A., Zelenika-Zovko, I., 2010. 3-D printing of open source appropriate technologies for self-directed sustainable development. J. Sustain. Develop 3 (4), 17.

Pearce, J.M., Grafman, L., Colledge, T., Legg, R., 2008. Leveraging information technology, social entrepreneurship and global collaboration for just sustainable development. Proceedings of the 12th NCIIA. Conference, 201–210.

Pretty, J.N., 1995. Regenerating Agriculture: Policies and Practice for Sustainability and Self-Reliance. Joseph Henry Press.

Reijntjes, C., Haverkort, B., Waters Bayer, A., 1992. Farming for the Future: An Introduction to Low-External-Input and Sustainable Agriculture. Macmillan.

Roling, N.G., Wagemakers, M.A.E. (Eds.), 2000. Facilitating Sustainable Agriculture: Participatory Learning and Adaptive Management in Times of Environmental Uncertainty. Cambridge University Press.

Sianipar, C.P., Yudoko, G., Dowaki, K., Adhiutama, A., 2013. Design methodology for appropriate technology: Engineering as if people mattered. Sustainability 5 (8), 3382–3425.

Tilman, D., Cassman, K.G., Matson, P.A., Naylor, R., Polasky, S., 2002. Agricultural sustainability and intensive production practices. Nature 418 (6898), 671–677.

United Nations Children's Fund (UNICEF), 2006. The state of the World's children. 148.

Weber, S., 2004. The Success of Open Source, Vol 368, Harvard University Press, Cambridge, MA.

Willinsky, J., 2006. The access principle: The case for open access to research and scholarship. MIT Press. http://arizona.openrepository.com/arizona/bitstream/10150/106529/1/jwapbook.pdf.

Zelenika, I., Pearce, J.M., 2011. Barriers to appropriate technology growth in sustainable development. J. Sustain. Develop 4 (6), 12.

Printed in the United States
By Bookmasters